Handbook of Dependability Engineering

Handbook of Dependability Engineering

Edited by
Maria Harris

MURPHY & MOORE
www.murphy-moorepublishing.com

Published by Murphy & Moore Publishing,
1 Rockefeller Plaza,
New York City, NY 10020, USA

ISBN: 978-1-63987-277-0

Cataloging-in-Publication Data

Handbook of dependability engineering / edited by Maria Harris.
p. cm.
Includes bibliographical references and index.
ISBN 978-1-63987-277-0
1. Reliability (Engineering). 2. Systems engineering. I. Harris, Maria.
TA169 .H36 2022
620.004 52--dc23

For information on all Murphy & Moore Publications
visit our website at www.murphy-moorepublishing.com

MURPHY & MOORE

Contents

Contents

Preface

The measure of a system's maintainability, availability, reliability and maintenance support performance is known as dependability. Some of its characteristics are safety, durability and security. Attributes, threats, and means are the major three elements of dependability. Attributes refer to the qualities of a system. Availability, reliability, safety, integrity and maintainability are dependability attributes. The objects or phenomenon which affect a system and result in a drop in its dependability are termed as threats. Means are constructed in order to break the Fault-Error-Chains so as to increase the dependability of a system. The four different means are prevention, removal, forecasting and tolerance. This book unravels the recent studies in the field of dependability engineering. It explores all the important aspects of this discipline in the present day scenario. This book will provide comprehensive knowledge to the readers.

Significant researches are present in this book. Intensive efforts have been employed by authors to make this book an outstanding discourse. This book contains the enlightening chapters which have been written on the basis of significant researches done by the experts.

Finally, I would also like to thank all the members involved in this book for being a team and meeting all the deadlines for the submission of their respective works. I would also like to thank my friends and family for being supportive in my efforts.

Editor

1

Modeling Quality of Service Techniques for Packet-Switched Networks

Wlodek M. Zuberek and Dariusz Strzeciwilk

Abstract

Quality of service is the ability to provide different priorities to different applications, users or dataflows, or to guarantee a certain level of performance to a dataflow. The chapter uses timed Petri nets to model techniques that provide the quality of service in packet-switched networks and illustrates the behavior of developed models by performance characteristics of simple examples. These performance characteristics are obtained by discrete-event simulation of analyzed models.

Keywords: quality of service, packet-switched networks, timed Petri nets, priority queuing, fair queuing, weighted fair queuing, performance analysis, discrete-event simulation

1. Introduction

Quality of service (or simply QoS) is the ability to provide different priorities to different applications, users or dataflows, or to guarantee a certain level of performance to a dataflow [1]. For example, a computer network can guarantee certain levels of error rate or jitter, or maximal delay of transmitted information. Quality of service guarantees are important if the network capacity is insufficient, especially for real-time streaming multimedia applications such as voice over internet (voice over IP), TV over internet (TV over IP), or multimedia applications, since these are often delay sensitive and require fixed bit rate [1, 2]. Similarly, quality of service is essential in networks where the capacity can be a limiting factor, for example, in cellular data communication [3].

There are two principal approaches to QoS in modern packet-switched IP networks, an integrated services approach based on application requirements that are exchanged with the network, and a differentiated approach where each packet identifies a desired service level to the network [4]. Early networks used the integrated services approach. It was realized, however, that in broadband networks, the core routers are required to deal with tens of thousands of service requirements—the integrated services approach does not scale well with the growth of the internet. Routers supporting differentiated services configure their network schedulers to use multiple queues for packets awaiting transmission. In practice, packets requiring low jitter (e.g., voice over IP or videoconferencing) are given priority over packets of other types. Typically, some bandwidth is also allocated to network control packets, which must be sent over the network without any unnecessary delay.

Modern packet-switched networks are complex structures [5], which, for modeling, require a flexible formalism that can easily handle concurrent activities as well as synchronization of different events and processes that occur in such networks. Petri nets [6, 7] are such formal models.

As formal models, Petri nets are bipartite directed graphs, in which two types of vertices represent, in a very general sense, conditions and events. An event can occur only when all conditions associated with it (represented by arcs directed to the event) are satisfied. An occurrence of an event usually satisfies some other conditions, indicated by arcs directed from the event. So, an occurrence of one event causes some other event (or events) to occur, and so on.

In order to study performance aspects of systems modeled by Petri nets, the durations of modeled activities must also be taken into account. This can be done in different ways, resulting in different types of temporal nets. In timed Petri nets [8], occurrence times are associated with events, and the events occur in real time (as opposed to instantaneous occurrences in other models). For timed nets with constant or exponentially distributed occurrence times, the state graph of a net is a Markov chain (or an embedded Markov chain), in which the stationary probabilities of states can be determined by standard methods [9]. These stationary probabilities are used for the derivation of many performance characteristics of the model [10].

In this chapter, timed Petri nets are used to model priority queuing systems, which provide the quality of service in packet-switched networks. Section 2 recalls basic concepts of Petri nets and timed Petri nets. Section 3 discusses (strict) priority queuing and its performance characteristics. Fair scheduling is described and illustrated in Section 4, while Section 5 deals with weighted fair scheduling. A combined approach, using several types of scheduling methods, is outlined in Section 6. Section 7 concludes the chapter.

2. Timed Petri nets

In Petri nets, concurrent activities are represented by *tokens,* which can move within a (static) graphlike structure of the net. More formally, a marked inhibitor place/transition Petri net $\mathcal{M}$ is defined as a pair $\mathcal{M} = (\mathcal{N}, m_0)$, where the structure $\mathcal{N}$ is a bipartite-directed graph, $\mathcal{N} = (P, T, A, H)$, with two types of vertices, a set of places P and a set of transitions T, a set of directed arcs A connecting places with transitions and transitions with places, $A \subseteq T \times P \cup P \times T$, and a set of inhibitor arcs H connecting places with transitions, $H \subset P \times T$; usually $A \cap H = \emptyset$. The initial marking function m_0 assigns non-negative numbers of tokens to places of the net, $m_0 : P \rightarrow \{0, 1, \ldots\}$. Marked nets can be equivalently defined as $\mathcal{M} = (P, T, A, H, m_0)$.

A place is shared if it is connected to more than one transition. A shared place p is free choice if the sets of places connected by directed arcs and inhibitor arcs to all transitions sharing p are identical. A shared place p is (dynamically) conflict free if for each marking reachable from the initial marking at most one transition sharing p is enabled. A net is a free choice if all its shared places are either free choice or (dynamically) conflict free. Only free-choice nets are used in this chapter.

In timed nets [8], occurrence times are associated with transitions, and transition occurrences are real-time events; i.e., tokens are removed from input places at the beginning of the occurrence period, and they are deposited to the output places at the end of this period. All occurrences of enabled transitions are initiated in the same instants of time in which the transitions become enabled (although some enabled transitions may not initiate their occurrences). If, during the occurrence period of a transition, the transition becomes enabled again, a new, independent occurrence can be initiated, which will overlap with the other occurrence(s). There is no limit on the number of simultaneous occurrences of the same transition (sometimes, this is called infinite occurrence semantics). Similarly, if a transition is enabled "several times" (i.e., it remains enabled after initiating an occurrence), it may start several independent occurrences in the same time instant.

More formally, a free-choice timed Petri net is a triple, $\mathcal{T} = (\mathcal{M}, c, f)$, where $\mathcal{M}$ is a marked net, c is a choice function that assigns probabilities to transitions in free-choice classes, $c \rightarrow [0, 1]$, and f is a timing function that assigns an (average) occurrence time to each transition of the net, $f: T \rightarrow \mathbf{R}^+$, where $\mathbf{R}^+$ is the set of non-negative real numbers.

The occurrence times of transitions can be either deterministic or stochastic (i.e., described by some probability distribution function); in the first case, the corresponding timed nets are referred to as D-timed nets [11]; in the second, for the (negative) exponential distribution of occurrence times, the nets are called M-timed nets (Markovian nets) [12]. In both cases, the concepts of state and state transitions have been formally defined and used in the derivation of different performance characteristics of the model. In simulation applications, other distributions can also be used; for example, the uniform distribution (U-timed nets) is sometimes a convenient option. In timed Petri nets, different distributions can be associated with different transitions in the same model providing flexibility that is used in simulation examples that follow.

In timed nets, the occurrence times of some transitions may be equal to zero, which means that the occurrences are instantaneous; all such transitions are called immediate (while the others are called timed). Since the immediate transitions have no tangible effects on the (timed) behavior of the model, it is convenient to "split" the set of transitions into two parts, the set of immediate and the set of timed transitions, and to first perform all occurrences of the (enabled) immediate transitions, and then (still in the same time instant), when no more immediate transitions are enabled, to start the occurrences of (enabled) timed transitions. It should be noted that such a convention effectively introduces the priority of immediate transitions over the timed ones, and therefore conflicts of timed and immediate transitions are not allowed in timed nets. Detailed characterization of the behavior or timed nets with immediate and timed transitions is given in [8].

3. Priority queuing

The basic idea of priority scheduling [13] is that separate queues are used by servers for packets of different priority classes, as shown in **Figure 1** (there are three classes of priorities with queues Q1, Q2, and Q3, for example, for voice, video and data packets, respectively). It is assumed that priorities decrease with the queue numbers; i.e., Q1 is the highest priority queue. In **Figure 1**, a_1 is the arrival rate of packets in priority class 1, a_2 is the arrival rate in class 2, and s is the service rate; $1/s$ is the average transmission time of a packet over the communication channel (represented as the server in **Figure 1**). For simplicity of the description, it is assumed that the transmission times are (approximately) the same for packets of different classes, but this can easily be replaced by rates s_1, s_2, and s_3.

To select a packet for transmission, the scheduler first checks the highest priority queue (Q1) and if this queue is nonempty, the scheduler uses the first packet from this queue. If Q1 is empty, the scheduler checks the next queue in the order of priorities (i.e., Q2 and then Q3). Consequently, a lowest priority packet is transmitted only when all other (higher priority) queues are empty.

Figure 2 shows a Petri net model of priority queuing outlined in **Figure 1**. Transitions t_{01}, t_{02}, and t_{03} with places p_{01}, p_{02}, and p_{03} are (independent) sources of packets for classes 1, 2, and 3,

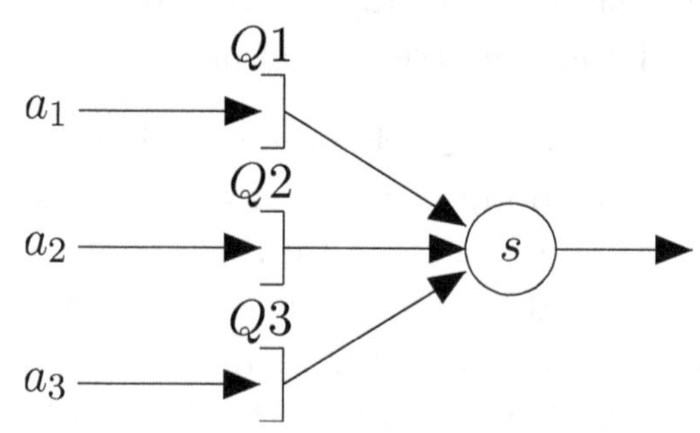

Figure 1. Priority queuing.

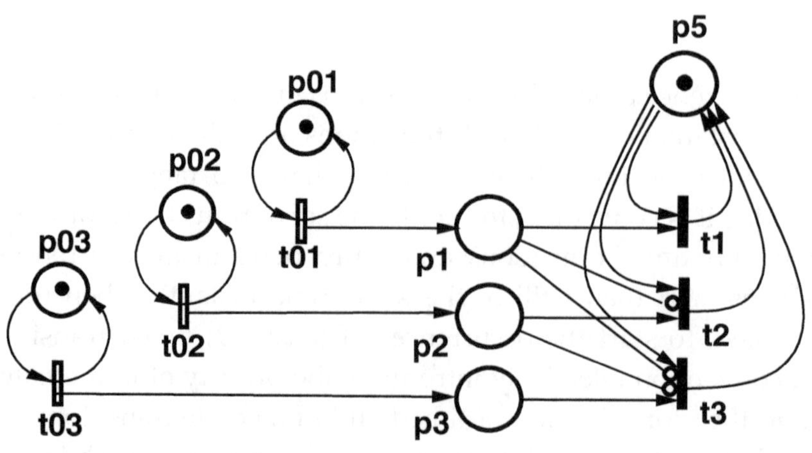

Figure 2. Timed Petri net model of priority queuing.

respectively. The rates of generated packets are determined by the occurrence times associated with t_{01}, t_{02}, and t_{03}; $a_1=1/f(t_{01})$, etc. Also, the mode of the timed transition (M, D, or U) determines the distribution of the interarrival times for the respective source.

Transitions t_1, t_2, and t_3 model the transmission times for packets of classes 1, 2, and 3, respectively. Place p_5, shared by these three transitions, guarantees that no two packets can be transmitted at the same time. Finally, the inhibitor arcs (p_1, t_2), (p_1, t_3), and (p_2, t_3) implement the priorities: if there is a packet of class 1 waiting for transmission (in p_1), no packet of class 2 or 3 can be transmitted; if there is a packet of class 1 or 2, no class 3 packet can be transmitted.

The priority scheme works well when the traffic intensity is small; when, however, the traffic becomes intensive, it is quite possible that bursts of packets of higher priorities can (temporarily) block the transmission of lower priority packets for extended periods of time, significantly degrading their performance. **Figure 3** shows the utilization of a transmission channel shared by three streams of packets (as in **Figure 1**) as a function of traffic intensity of packets of priority 1, ρ_1, with fixed traffic intensities for packets of priorities 2 and 3 (at the values of $\rho_2=0.5$ and $\rho_3=0.25$).

It can be observed that for traffic intensity $\rho_1>0.25$, i.e., $\rho_1>1.0-\rho_2-\rho_3$, channel utilization for packets of priority 3 decreases to zero; for $\rho_1\geq 0.5$, packets of priority 3 are blocked. Similarly, for $\rho_1>0.5$, the utilization of the channel for packets of priority 2 decreases, which means that only some of such packets can be send through the channel.

Waiting times of packets with priorities 1, 2, and 3, for the case shown in **Figure 3**, are presented in **Figure 4**. As the traffic intensity ρ_1 approaches 0.25, waiting times for packets of priority 3 increase indefinitely, and for ρ_1 approaching 0.5, so do waiting times of packets of priority 2.

It should be noted that for $\rho_1>0.25$, queue Q3 (with infinite capacity) is nonstationary as the rate of packets entering the queue is greater than the rate of packets removed from it (for transmission through the channel). Similarly, queue Q2 (with infinite capacity) is also nonstationary for

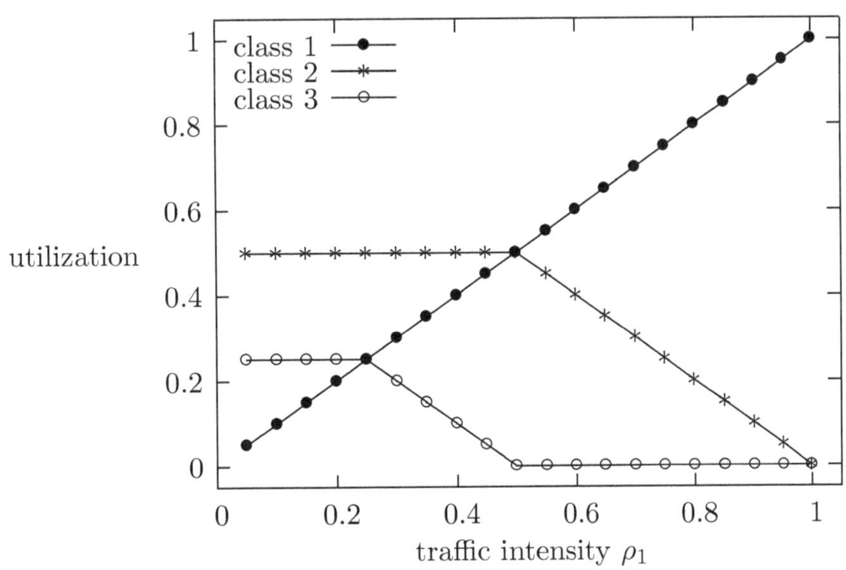

Figure 3. Channel utilization as a function of ρ_1 with $\rho_2=0.5$ and $\rho_3=0.25$.

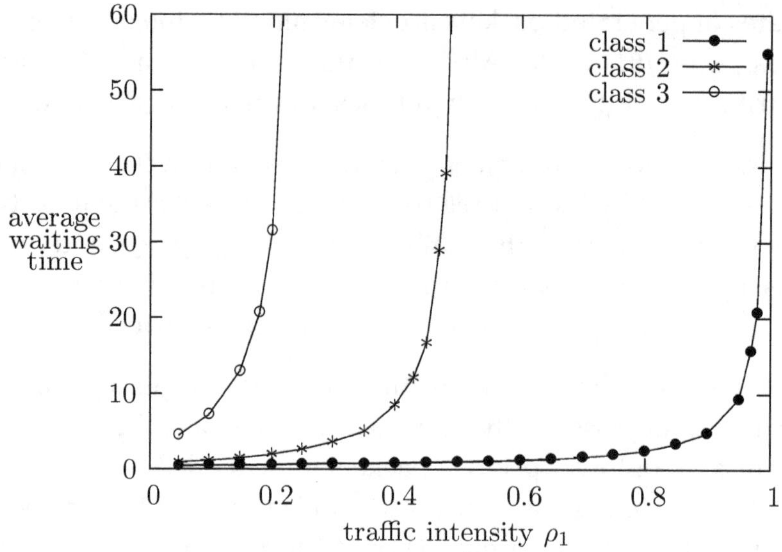

Figure 4. Average waiting times as functions of traffic intensity ρ_1 with ρ_2=0.5 and ρ_3=0.25.

ρ_1>0.5. This is the reason that, in **Figure 4**, waiting times are shown only for stationary regions of behavior.

4. Fair queuing

In fair queuing [14], a transmission medium is shared (in a fair way, i.e., in "equal parts") by all classes of traffic (and the classes of traffic correspond to different packet flows, e.g., voice over IP, video, etc.). Implementations of fair queuing use separate queues for different classes of traffic, and packets are forwarded from these queue in a cyclic way, providing the same service for all classes.

Timed Petri nets model of fair queuing with three classes of traffic (as in **Figure 1**) is shown in **Figure 5**.

Places p_1, p_2, and p_3 are queues for classes 1, 2, and 3, respectively. If the queues for all classes are nonempty, the selection is performed cyclically in a loop:

$$p_{31}, t_{11}, p_{10}, t_{10}, p_{12}, t_{22}, p_{20}, t_{20}, p_{23}, t_{33}, p_{30}, t_{30}, p_{31}.$$

This loop is modified if (at the selection time) some queues are empty. For example, if after selecting a packet of class 1 packet, the queue for class 2 is empty, while the queue for class 3 is nonempty, the sequence is:

$$p_{31}, t_{11}, p_{10}, t_{10}, p_{12}, t_{32}, p_{30}, t_{30}, p_{21}.$$

If the only nonempty queue is the queue for class 1, the sequence is

$$p_{31}, t_{12}, p_{10}, t_{10}, p_{31}.$$

Generally, there are three cases when packets are selected from (nonempty) queue 1:

- queue 1 is checked in the order of fair queuing (marked place p_{31}) and the queue is nonempty (i.e., transition t_{11} can occur),
- queue 3 is checked in the order of fair queuing (marked place p_{23}), but the queue is empty so queue 1 is checked as the next one and it is nonempty (i.e., transition t_{13} can occur),
- queue 2 is checked in the order of fair queuing, but queue 2 as well as the next queue, queue 3, are empty, while queue 1 is nonempty (i.e., transition t_{12} can occur).

There are similar cases for queues 2 and 3.

It should be observed, that the set of places:

$$\{p_{31}, p_{12}, p_{23}, p_{10}, p_{20}, p_{30}\}$$

always contains a single token (initially shown in p_{31} in **Figure 5**). The cyclic checking of queues of fair queuing is controlled by this token.

Figure 6 shows the average waiting times for a system with fair queuing and three classes of traffic when the traffic intensities are the same for all three classes. Since the service rates are

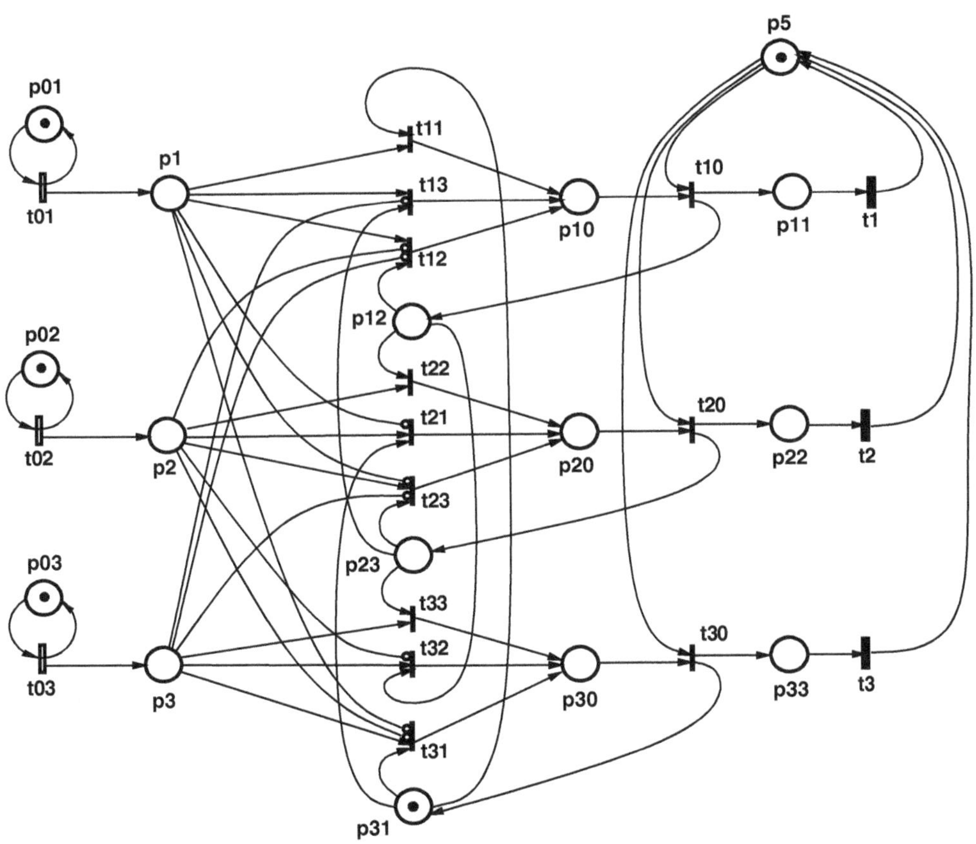

Figure 5. Timed Petri net model of fair queuing.

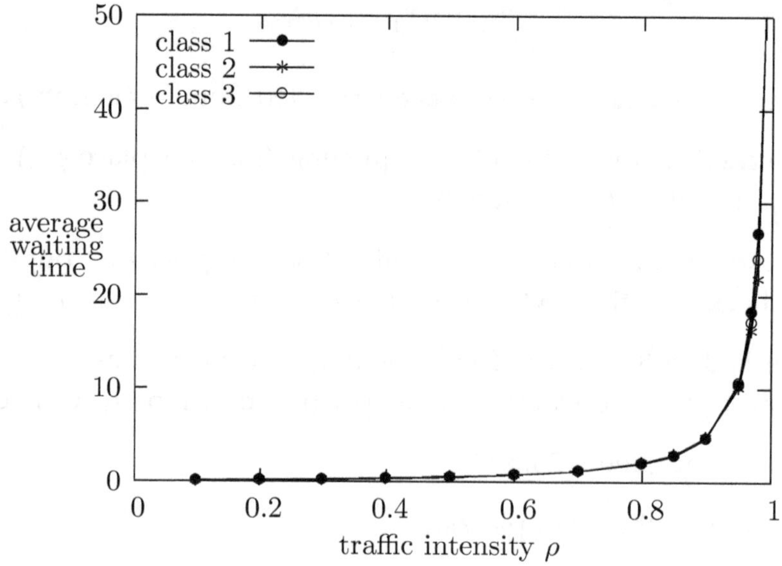

Figure 6. Average waiting times as functions of traffic intensity ρ with $\rho_1 = \rho_2 = \rho_3$.

also the same for all classes, channel utilizations as well the average waiting times for are the same in this case for all classes of traffic.

If, however, traffic intensities are different for different classes, average waiting times as well as utilizations of the shared transmission channel are also different. **Figure 7** shows the average waiting times for the case when $\rho_1 = 0.5\rho$, $\rho_2 = \rho_3 = 0.25\rho$.

In this case, the average waiting times for traffic classes with greater traffic intensity (class 1) increase much faster with increasing traffic intensity than for other classes of traffic (classes 2 and 3).

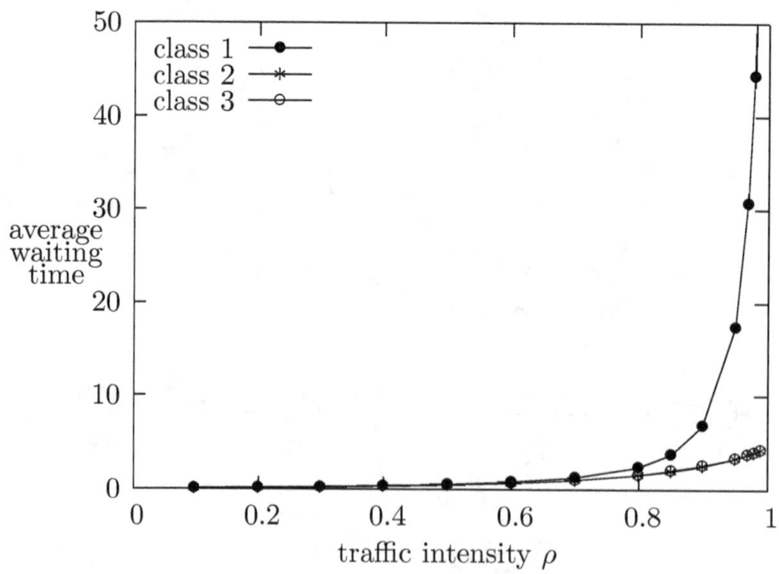

Figure 7. Average waiting times as functions of traffic intensity ρ with $\rho_1 = 0.5\rho$, $\rho_2 = \rho_3 = 0.25\rho$.

5. Weighted fair queuing

Weighted fair scheduling [13] restricts the priority scheduling by introducing limits on the number of consecutive packets of the same class that can be transmitted over the channel; when the scheduler reaches such a limit, it switches to the next nonempty priority queue and follows the same rule. So, if there are sufficient supplies of packets in all priority classes, the scheduler selects w_1 packets of class 1, then w_2 packets of class 2, then w_3 packets of class 3, and again w_1 packets of class 1, and so on, where w_1, w_2, and w_3 are the weights for classes 1, 2 and 3, respectively. Consequently, in such a situation (i.e., for sufficient supply of packets in all classes), the channel is shared by the packets of all priority classes, and the proportions are

$$u_i = \frac{w_i/s_i}{\sum_{j=1,\ldots,k} w_j/s_j}, i = 1, 2, \ldots k, \quad (1)$$

where k is the number of priority classes and s_i, $i = 1, \ldots, k$, is the transmission rate for packets of class i. If the transmission rates are the same for packets of all priority classes (as is assumed for simplicity in the illustrating examples), the properties are

$$u_i = \frac{w_i}{\sum_{j=1,\ldots,k} w_j}, i = 1, \ldots, k. \quad (2)$$

For an example with three priority classes and the weights equal to 4, 2, and 1 for classes 1, 2, and 3, respectively, these "utilizations bounds" are equal to 4/7, 2/7, and 1/7, for classes 1, 2, and 3, respectively.

A Petri net model of weighted fair scheduling for three priority classes with weights 4, 2, and 1 is shown in **Figure 8**. The model is composed of three identical interconnected sections corresponding to the three priority classes. The main elements of the model are the three queues represented by places p_1, p_2, and p_3 for classes 1, 2 and 3, respectively, and timed transitions t_1, t_2, and t_3 modeling the transmission of selected packets through the communication channel. The three classes of packets are generated (independently) by transitions t_{01}, t_{02}, and t_{03} with places p_{01}, p_{02}, and p_{03}.

As in fair queuing, the scheduling is based on cyclic selection of queues for the transmission of waiting packets. This cyclic operation is represented by a (rather complex) loop with places r_1, r_2, and r_3; q_1, q_2, and q_3; and also s_1, s_2, and s_3. There is a single "control token" in this loop (shown in place s_3 in **Figure 8**). This token always indicates the queue that is used for transmission of packets.

The section for class 2 is shown separately in **Figure 9** to make its description easier to follow.

Place r_2 becomes marked only when nonempty queue 2 is used for the selection of packets. Place w_2 contains the weight of class 2 (in this case 2). Transition a_2 selects a packet (from p_2) and forwards it to p_{20}, moving a single token from w_2 to u_2. When the channel becomes available (i.e., p_5 becomes marked), the selected packet is forwarded from p_{20} to p_{22} and then is transmitted (transition t_2). At the same time, a token is returned by t_{20} to r_2 to allow selecting another packet from p_2. This is repeated until:

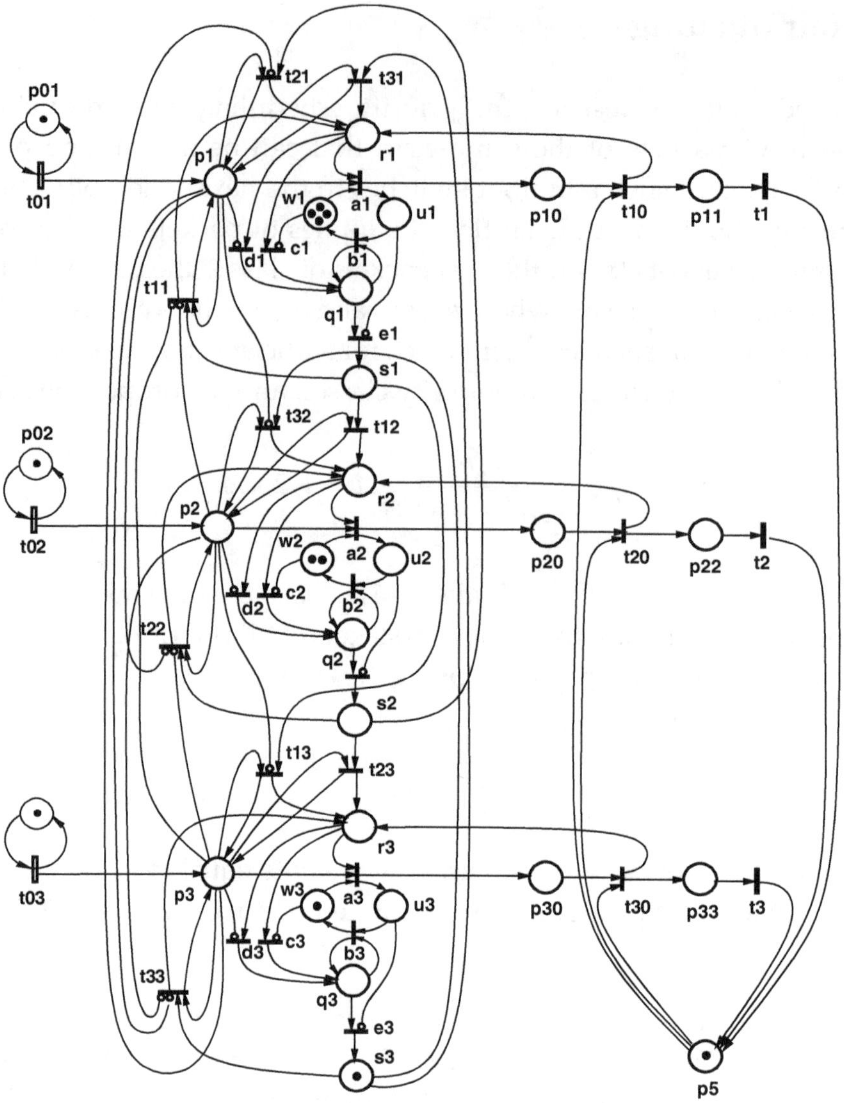

Figure 8. Petri net model of weighted fair queuing with weights 4-2-1.

- there are no more token in w_2 (transition c_2 occurs), or
- there are no more token in p_2 (transition d_2 occurs).

In both cases, the token from r_2 is moved to q_2 and then a number of occurrences of b_2 moves all tokens form u_2 back to w_2. When u_2 becomes unmarked, transition e_2 moves the token from q_2 to s_2 in order to select the next traffic class. If p_3 is nonempty, transition t_{23} moves the token from s_2 to r_3. If p_3 is unmarked and p_1 is marked, transition t_{21} moves the token from s_2 to r_1. If both p_1 and p_3 are unmarked but p_2 is marked, transition t_{22} moves the token from s_2 to r_2 and transmission of class 2 packets continues. Finally, if none of t_{21}, t_{22}, and t_{23} is enabled, the token remains in s_2 waiting for a packet arriving to p_1, p_2, or p_3.

It should be observed that when the traffic is intense, i.e., when the queues p_1, p_2, and p_3 are nonempty most of the time, the queuing mechanism repeatedly selects w_1 packets from p_1, then w_2 packets from p_2, then w_3 packets from p_3, and so on. On the other hand, when the traffic is light, all packets are transmitted with very little delay.

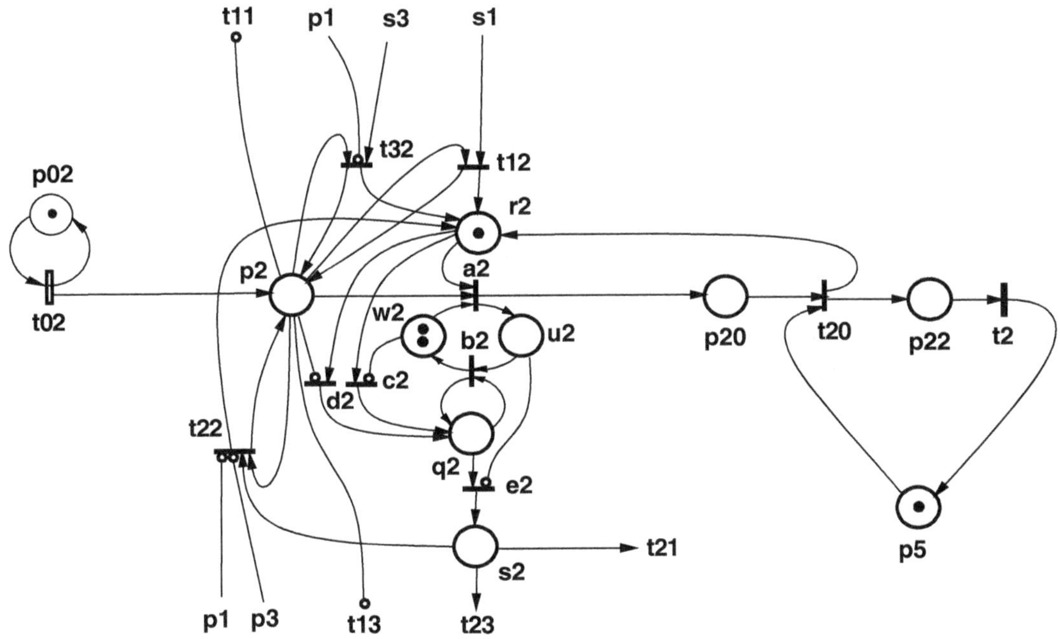

Figure 9. Petri net model of class 2 of weighted fair queuing.

Figure 10 shows the utilization of the transmission channel shared by three classes of packets (as in **Figure 1**) as a function of traffic intensity of traffic class 1, ρ_1, with fixed traffic intensities for traffic classes 2 and 3 (at the values of $\rho_2=0.5$ and $\rho_3=0.25$).

For $\rho_1>0.25$, channel utilization for packets of priority 3 decreases from the initial value of 0.25 to its weighted value of 0.14 (i.e., 1/7). Also, channel utilization for packets of priority 2 decreases from its initial value of 0.5 to its weighted value of 0.28 (i.e., 2/7). At $\rho_1=0.57$ (i.e., 4/7), the total utilization of the channel becomes 100% and no further increase of utilization is possible.

Similarly as before (**Figures 3** and **4**), queues Q2 and Q3 are nonstationary for $\rho_1>0.25$ and queue Q1 is nonstationary for $\rho_1>0.57$ (i.e., 4/7).

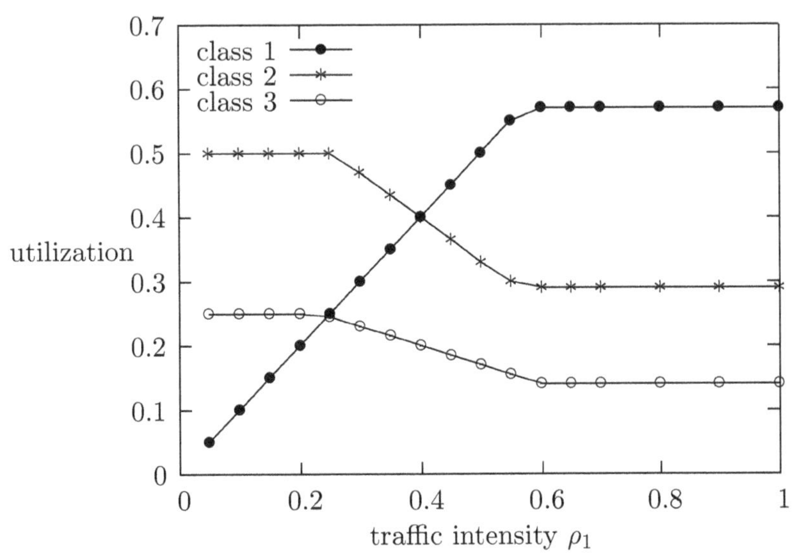

Figure 10. Channel utilization as a function of ρ_1 with $\rho_2=0.5$ and $\rho_3=0.25$.

For the same weights but for different (fixed) arrival rates for classes 2 and 3, i.e., for $\rho_2=0.25$ and $\rho_3=0.1$, the utilization of the transmission channel as a function of traffic intensity of traffic class 1 is shown in **Figure 11**.

For this case, queues Q2 and Q3 are stationary for $0 \le \rho_1 < 1$ and Q3 is nonstationary for $\rho_1 > 0.65$ (i.e., 1.0–0.25–0.1). The average waiting times for classes 2 and 3 depend in a limited way on the traffic intensity ρ_1, as shown in **Figure 12**.

In weighted fair queuing, if weights are equal to the arrival rates, the traffic of lower priority classes is (almost) independent of the traffic intensity of higher-priority classes; the effect is

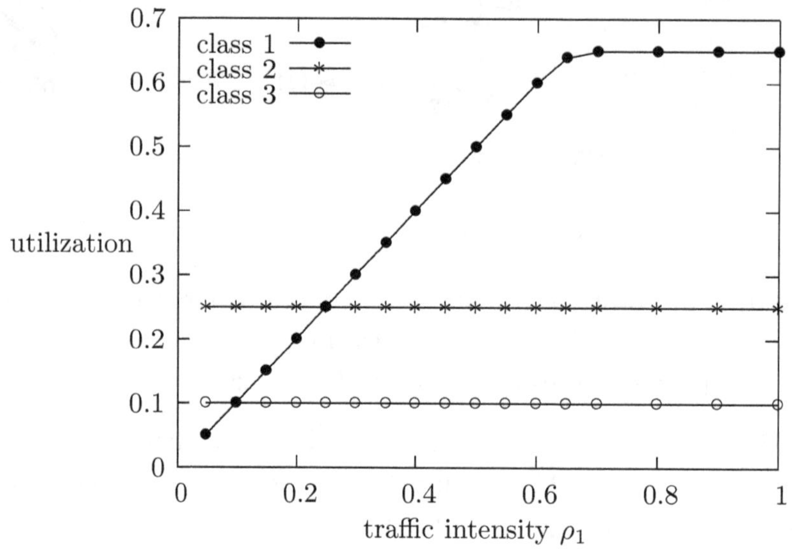

Figure 11. Channel utilization as a function of ρ_1 with $\rho_2=0.25$ and $\rho_3=0.1$.

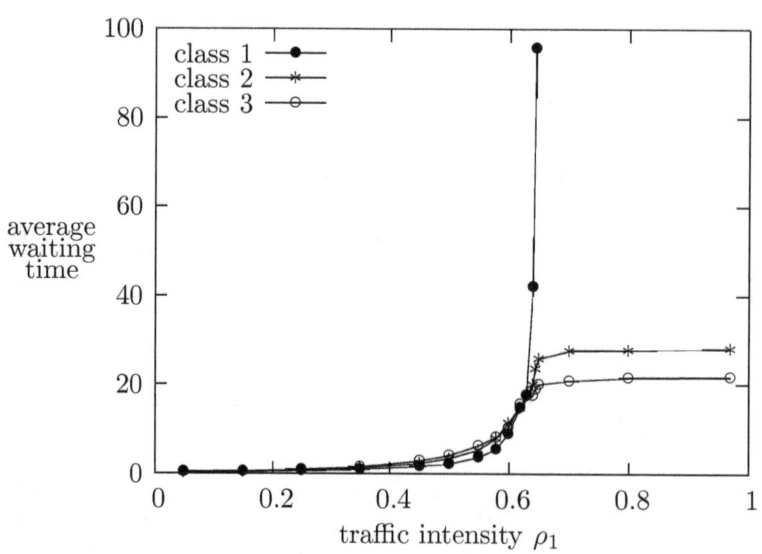

Figure 12. Average waiting times as functions of ρ_1 with $\rho_2=0.25$ and $\rho_3=0.1$.

similar to assigning some (shared) transmission capacity to lower priority traffic classes. Moreover, if this "reserved" capacity is not used, it is available to other classes of traffic.

6. Combined queuing

The basic queuing methods discussed earlier can be combined into more complex systems. For example, the combination of priority queuing and weighted fair queuing is known as low-latency queuing (LLQ) [15]. A simpler case of combining priority queuing and fair queuing, as shown in **Figure 13**, is used as an illustration of the combined approach.

It is assumed in this example that 40% of bandwidth is allocated to the priority queue (Q1) and that the remaining 60% of bandwidth is equally divided among queues Q2, Q3 and Q4.

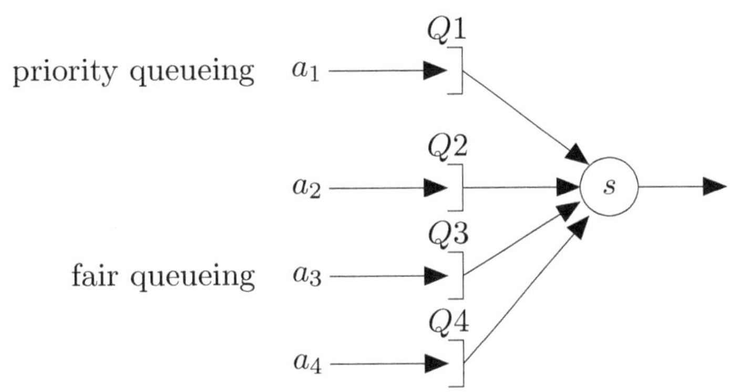

Figure 13. Priority queuing combined with fair queuing.

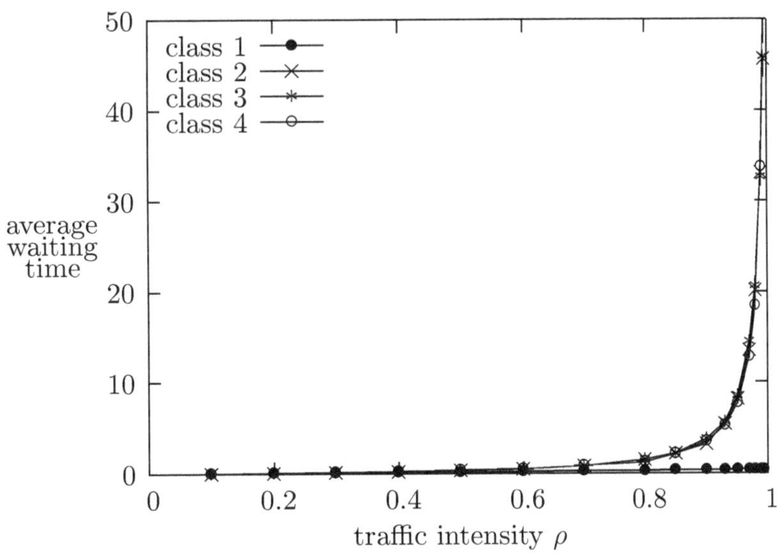

Figure 14. Average waiting times as functions of ρ with $\rho_1 = 0.4\rho$ and $\rho_2 = \rho_3 = \rho_4 = 0.2\rho$.

Average waiting times for all four classes of traffic (**Figure 13**) as functions of traffic intensity are shown in **Figure 14**.

Figure 14 shows the effects of priority scheduling (class 1) on lower priority classes (classes 2, 3, and 4) when traffic intensity approaches 1—the average waiting times increase rather significantly for classes 2, 3, and 4 while class 1 remains practically unaffected by the increased traffic. Also, because of fair queuing, the average waiting times for classes 2, 3, and 4 are practically identical.

7. Concluding remarks

The Internet 2 project, launched in 2001, was probably too early for implementation of QoS protocols with the equipment that was then available [16]. It should not be surprising that this resulted in a conclusion that adding more bandwidth (i.e., over-provisioning) is more effective than any of the various schemes for accomplishing QoS [17]. But cost and other factors prevent service providers to built and maintain permanently over-provisioned networks; other approaches must be used to guarantee the performance of services available in modern packet-switched networks [16].

Several models of techniques used for providing quality of service in packet-switched networks are discussed in this chapter. These models are used to derive performance characteristics of scheduling methods and to provide some insights into the behavior of packet-switched networks. In particular, the blocking of lower priority classes of traffic, typical for priority scheduling, can easily be observed. Also, the guaranteed levels of service of fair scheduling and weighted fair scheduling can easily be illustrated.

It should be noted, however, that the discussed techniques are just basic elements of complex computer networks and that the behavior of real systems is very dynamic and usually difficult to predict [18]. Therefore, more work is needed in this area to use the networks in an efficient and predictable way.

Also, an attractive aspect of the models would be some kind of compositionality that would allow models to be easily combined into more complex ones, as outlined in Section 6. The models presented in this chapter need to be revised to make such compositions straightforward.

Author details

Wlodek M. Zuberek[1]* and Dariusz Strzeciwilk[2]

*Address all correspondence to: wlodek@mun.ca

1 Department of Computer Science, Memorial University, St. John's, NL, Canada

2 Department of Applied Informatics, University of Life Sciences, Warszawa, Poland

References

[1] Guerin R, Peris V. Quality of service in packet networks: Basic mechanisms and directions. Computer Networks. 1999;**31**:169-189

[2] Wang Z. Internet QoS: Architectures and Mechanisms for Quality of Service. San Francisco, CA: Morgan Kaufmann; 2001

[3] Marchese M. QoS over Heterogeneous Networks. Chichester: Wiley and Sons; 2007

[4] Fergusson P, Huston G. Quality of Service: Delivering QoS on the Internet and in Corporate Networks. New York, NY: Wiley and Sons; 1998

[5] Robertazzi TG. Computer Networks and Systems: Queueing Theory and Performance Evaluation. Berlin, Heidelberg: Springer-Verlag; 1990

[6] Murata T. Petri nets: Properties, analysis and applications. Proceedings of IEEE. 1989;**77**(4): 541-580

[7] Reisig W. Petri Nets — An Introduction (EATCS Monographs on Theoretical Computer Science 4). Berlin, Heidelberg: Springer-Verlag; 1985

[8] Zuberek WM. Timed Petri nets — Definitions, properties and applications. Microelectronics and Reliability (Special Issue on Petri Nets and Related graph models). 1991;**31**(4): 627-644

[9] Allen AA. Probability, Statistics and Queueing Theory with Computer Science Applications. 2nd ed. San Diego, CA: Academic Press; 1991

[10] Jain R. The Art of Computer Systems Performance Analysis. Berlin, Heidelberg: Springer-Verlag; 1991

[11] Zuberek WM. D-timed Petri nets and modelling of timeouts and protocols. Transactions of the Society for Computer Simulation. 1987;**4**(4):331-357

[12] Zuberek WM. M-timed Petri nets, priorities, preemptions, and performance evaluation of systems. In: Advances in Petri Nets 1985 (LNCS 222). Berlin, Heidelberg: Springer-Verlag; 1986. p. 478-498

[13] Georges P, Divoux T, Rondeau E. Strict priority versus weighted fair queueing in switched Ethernet networks for time-critical applications. In: Proc. 19-th IEEE International Parallel and Distributed Processing Symposium (IPDPS'05); 2005. pp. 141-145

[14] Park KI. QoS in Packet Networks. Boston, MA: Springer Science; 2005

[15] Dekeris B, Adomkus T, Budnikas A. Analysis of QoS assurance using weighted fair queueing (WFQ) scheduling with low latency queue (LLQ). In: Proceedings of 26-th Int. Conference on Information Technology Interfaces (ITI'06); 2006. pp. 507-512

[16] Lindgren A, Almquist A, Schelen O. Quality of service for IEEE 802.11—A simulation study. In: Quality of Service – IWQoS 2001 (LNCS 2092). Berlin, Heidelberg: Springer-Verlag; 2001. p. 281-287

2

Time Series and Renewable Energy Forecasting

Mahmoud Ghofrani and Anthony Suherli

Abstract

Reliability is a key important criterion in every single system in the world, and it is not different in engineering. Reliability in power systems or electric grids can be generally defined as the availability time (capable of fully supplying the demand) of the system compared to the amount of time it is unavailable (incapable of supplying the demand). For systems with high uncertainties, such as renewable energy based power systems, achieving a high level of reliability is a formidable challenge due to the increased penetrations of the intermittent renewable sources such as wind and solar. A careful and accurate planning is at the utmost importance to achieve high reliability in renewable energy based systems. This chapter will assess wind-based power system's reliability issues, and provide a case study that proposes a solution to enhance the reliability of the system.

Keywords: availability, energy storage, renewable energy, reliability, wind

1. Introduction

The world is moving forward in technology as power systems lean toward renewable energy more and more each year. While the idea of using renewable energy has long been the focus of numerous researches from all over the world, the implementation itself is more complicated than said. Dealing with renewable energy proposes new challenges that must be carefully addressed and solved. The uncertainty of renewable energy sources, such as wind speed (for wind turbines) or solar radiation (for solar photovoltaic (PV) panels), and the fact that it is unreliable from time to time due to said uncertainty, are two of the major issues that rough up the transition from fossil based energy sources to renewable energy sources. The main objective of operational and planning strategies is to enable power systems to constantly and continuously meet the consumers' demand or the system load. The volatility of renewable energy sources jeopardizes the power system's ability to reliably meet this objective.

Climate change concerns, and new state initiatives are some of the factors that contributed in pushing and escalating the number of wind power based technology deployment during recent years. The stochastic nature of wind power resources makes it difficult to perform a reliable operation. While this issue has been frequently studied and numerous methods have been developed, a flawless solution for every case has not yet been designed. One of the proposed solutions is to use fast-responding units like gas generators as the operating reserves to keep up with the demand [1, 2], although doing that reduces the system's efficiency and increases its operating costs [3]. Another possible solution is to install energy storage systems, which store wind power during low-demand periods and release power during periods when the system cannot provide sufficient power to meet the load [4]. This increases the flexibility of the power system as the energy storage system counterbalances the unexpected wind power fluctuations to more efficiently utilize the smoothened wind power for supplying the system demand.

In the upcoming future, energy storage systems are expected to be an essential part of electric grids. However, its deployment depends heavily on its economic advantages when compared to the more conventional operational practices. To come up with the most economically beneficial plan, a cost–benefit analysis must be done for each possible technology, especially in regulated utilities where the limited market opportunities diminish the potential economic benefits of storage technologies over gas-fired generators [3]. To assure the effectiveness of an energy storage system, we must approach the problem with an appropriate strategy [5, 6]. An optimal storage sizing strategy furnishes the system with the capability to stabilize against forecast uncertainty and integrate wind power more reliably [7–9], added with optimal scheduling, it also improves the system's transmission capacity utilization.

For high wind penetrations, fast-response thermal units are used as reserve capacities to provide the fast ramping capability required to deal with wind power fluctuations. Recent developments in storage technologies have advanced its energy efficiency and enhanced its capability in dealing with fast ramping. Additionally, storage systems bring forth several benefits when compared to fast-response thermal units, such as efficiency enhancement of renewable integration, reduced emission, and improved utilization of grid assets.

Several applications have been proposed for energy storage systems, which include but are not limited to renewable capacity firming and reliability enhancement of renewable integration. Each application requires a case-by-case optimal allocation strategy, which might result in different solutions. Particularly, the matter of optimally sizing, siting, scheduling and operating storage systems to address the reliability issues of intermittent renewable integration is of great importance. A solid, probabilistic optimization framework is needed to supplement grid operability and reliability while at the same time reduces overall costs for systems with high wind penetrations. The framework developed by the author in Ref. [10] is adopted for this chapter and is provided in the next section. A case study is presented in this chapter to analyze the reliability of renewable energy based systems and compare storage technologies and conventional gas-fired alternatives for reliably integrating different wind penetrations. An economic analysis is also provided to calculate costs and benefits associated with each technology to determine the most economical solution.

2. Methodology

The following methodology is one possible solution example to model an intermittent renewable energy-based power system.

2.1. Wind, load, and equipment availability modeling

We use probability distribution functions (PDFs) to model the stochastic nature of load and wind generation, which parameters are calculated using 10 years of historical hourly data for load and wind speed [11]. The produced model will then be used to generate hourly samples for the planning period. We use Fuzzy C-Means (FCM) clustering to capture a statistical model that takes into account seasonal variations [12]. We grade each of the sample points with a value within the range of [0, 1], then we minimize the weighted distance between any sample point and a cluster center by using an iterative algorithm. The elbow method determines the total number of clusters [13]. By combining the FCM clustering and the elbow method, we categorize our planning days into 40 clusters of 24-hour wind speed and load samples. We utilize the maximum likelihood method to find the parameters of the PDFs for the samples. Two sets of 24 individual PDFs will represent each of the clusters for a 24-hour period.

2.1.1. Wind power modeling

The total power generated by a wind turbine can be calculated by the product of a simple kinetic energy equation through a cross sectional area A as follows [14].

$$P = \frac{1}{2}\rho v^3 A = \frac{1}{2}\rho v^3 \frac{\pi d^2}{4} \tag{1}$$

where v is the wind speed in meters per second (m/s), ρ represents air density in kg/m^3 and d is the rotor diameter in meters.

The power output of a wind turbine depends heavily on the speed of the wind, and the wind speed itself can be best characterized by using the Weibull PDF, which formula is the following [11, 15]:

$$f(v; c, k) = \frac{k}{c}\left(\frac{v}{c}\right)^{k-1} e^{-\left(\frac{v}{c}\right)^k} \tag{2}$$

where k is a shape vector, c is the scale vector, and v is a vector of the measured wind speed. The average width of the wind speed distribution is determined by the shape vector k, while the scale vector indicates where the majority of the distribution lies and how wide the distribution-stretch is.

The wind power output can be calculated by using the power-speed curve [16]:

$$G_W = \begin{cases} 0 & v \le v_i, v \ge v_o \\ \frac{v - v_i}{v_r - v_i} G_{W_r} & v_i \le v \le v_r \\ G_{W_r} & v_r \le v \le v_o \end{cases} \tag{3}$$

where G_W is the output wind power, and v_i, v_0, v_r, v represents cut-in speed, cut-out speed, rated speed, and wind speed respectively.

2.1.2. Load modeling

The variation of the load is described by the Gaussian distribution [11]:

$$f(l; \mu, \sigma) = \frac{1}{\sqrt{2\pi\sigma^2}} \exp\left[-\frac{(L - \mu)^2}{2\sigma^2}\right] \tag{4}$$

where σ and μ represent the standard distribution, and mean of the Gaussian distribution respectively, and L represents the load demand.

2.1.3. Equipment availability modeling

Forced outage rate (*FOR*) of an equipment is the unavailability of the equipment estimated for a long-time period [17]. *FOR* models the availability of the equipment stochastically by the binomial PDF as follows:

$$f(q; n, p) = \binom{n}{q} p^q (1 - p)^{n-q} \tag{5}$$

where n is the number of units for each power plant and q = 0, 1, 2, …., n. The availability of each unit p is:

$$p = 1 - FOR \tag{6}$$

where *FOR* is basically the probability of the system's unavailability. For systems with long operating cycles, *FOR* can adequately estimate the unavailability probability of units that operates under similar conditions. On the other hand, it is not an adequate estimator for systems with short demand cycles. The most important period in the operation of a unit is the start-up period, and a peaking unit (example of a system with short demand cycles) will have less operating hours with more start-up and shut-down periods [17].

2.2. Energy storage modeling

The model of a storage system must be able to handle the energy balance between the sum of the stored and generated energy and the load, where it stores excess energy gained from wind generation and releases the energy to supply the peak demand. We can use compressed air energy storage (CAES) to enhance the wind integration performance in a transmission network due to its beneficial features such as large power capacity, long lifetime, and low operation costs [18]. The charging and discharging equations of the storage system are as follows:

$$S_t = (1 - d_s)S_{t-1} + \eta_{cs} L_{s_t} \qquad \forall t \in T \tag{7}$$

$$S_t = (1 - d_s)S_{t-1} - \eta_{ds} G_{s_t} \qquad \forall t \in T \tag{8}$$

where S_t represents the energy stored in the storage system at hour t, η_{cs} and η_{ds} represents the charging and discharging efficiencies for the CAES, L_{s_t} represents the storage loading capacity at hour t, and G_{s_t} represents the storage generating capacity at hour t, and d_s represents the self-discharge rate for CAES.

The state of charge of the storage system at any time t is within the minimum and maximum storage capacity requirements:

$$S_{min} \leq S_t \leq S_{max} \qquad \forall t \in T \tag{9}$$

where S_{min} and S_{max} are the minimum and maximum storage capacities.

The stored power must not exceed the maximum power rating at any given time as follows:

$$|P_t| \leq P_{max} \forall t \in T \tag{10}$$

where P_t and P_{max} are the storage power at time t and the maximum storage power respectively.

The following is the ramping constraints for the storage:

$$G_{S_t} - G_{S_{t-1}} \leq RU_s \qquad \forall t \in T \tag{11}$$

$$G_{S_{t-1}} - G_{S_t} \leq RD_s \qquad \forall t \in T \tag{12}$$

where RU_s and RD_s are the ramp up and ramp down of the turbine for the storage system respectively.

CAES has an expected lifetime of 30 years [19].

2.3. Economic modeling

The storage cost is the sum of the energy and power costs associated with each energy storage technology. The storage cost is described by the following equation [20, 21]:

$$IC_S = C_S.S_{max} + C_P.P_{max} \tag{13}$$

where IC_S is the cost of investment for the storage system, C_S is the energy cost for the storage system, and C_P is the power cost for the storage system.

The energy cost for CAES is 53 \$/kWh, which includes the combined reservoir and the balance of plant costs. The power cost of CAES is around 425 \$/kW [20], which includes turbine, compressor, and other power related costs.

The operation expenses are the sum of operation and maintenance (O&M) and fuel costs, which can be described by the following equation [22]:

$$OC_{S_t} = HR.G_{S_t}.C_{NG_t} + C_{OM}.P_{max} \qquad \forall t \in T \tag{14}$$

where OC_{S_t} is the operation cost of the storage system, HR is the turbine heat rate for the storage system, C_{NG_t} is the cost of natural gas of the storage system, and C_{OM} represents the cost of operation and maintenance for the storage.

C_{NG_t}, C_{OM}, and HR are 4300 Btu/kWh, 5 \$/MBtu and 2.5 \$/kW-year [20, 22].

For a gas-fired conventional generator, the investment cost and heat rate are 695 \$/kW and 8000 Btu/kWh respectively.

The total annual cost can be calculated by uniformly distributing the investment costs over the lifetime as follows:

$$A = \frac{d(1+d)^N}{(1+d)^{N+1} - 1} \cdot IC \tag{15}$$

where A is the annual equivalent cost for the investment, d is the discount rate, N is the life cycle of the investment, and IC is the investment cost.

We assume a discount rate of 10% and a lifetime of 30 years for the investment.

2.4. DC optimal power flow

We use optimal power flow (OPF) to find the steady state condition that at the same time minimizes the total operation and reliability costs. The objective function of the deterministic OPF is as follows:

$$\begin{aligned} Obj.Function = \text{Min}\left\{ \sum_{i=1}^{n_g} \left(a_i P_{g_{i,t}}{}^2 + b_i P_{g_{i,t}} + c_i \right) \right. \\ \left. + \sum_{i=1}^{n_b} IEAR_i \times IL_{i,t} \right\} \\ = \text{Min}(OC_t + ILC_t) \qquad \forall t \in T \end{aligned} \tag{16}$$

where $IEAR_i$ is the interrupted energy assessment rate at each bus. a_i, b_i and c_i are the coefficients of the cost function for the ith generator, $P_{g_{i,t}}$ represents the power output of the i-th generator at hour t, and ILC is the interrupted load cost.

The objective function above is subject to each of the following constraints:

Power balance equation:

$$\sum_{i=1}^{n_b} P_{g_{i,t}} = \sum_{i=1}^{n_b} P_{d_{i,t}} \qquad \forall t \in T \tag{17}$$

where $P_{d_{i,t}}$ is the supplied load at bus i at hour t, and n_b is the bus number.

Power generation and load limitations:

$$P_{g_{i,t}}^{min} \le P_{g_{i,t}} \le P_{g_{i,t}}^{max} \qquad \forall t \in T \tag{18}$$

$$P_{g_{i,t}}^{min} \le P_{g_{i,t}} \le P_{g_{i,t}}^{max} \qquad \forall t \in T \tag{19}$$

where $P_{g_{i,t}}^{min}$ and $P_{g_{i,t}}^{max}$ are the lower and upper generation limits for the i-th generator at hour t respectively.

Interrupted load:

$$IL_{i,t} = P_{D_{i,t}} - P_{d_{i,t}} \qquad \forall t \in T \tag{20}$$

where $P_{D_{i,t}}$ is the load demand at bus i at hour t.

Generation ramp up and ramp down:

$$P_{g_{i,t}} - P_{g_{i,t-1}} \le RU_i \qquad \forall t \in T \tag{21}$$

$$P_{g_{i,t-1}} - P_{g_{i,t}} \le RD_i \qquad \forall t \in T \tag{22}$$

And the transmission line limitation:

$$\sum_{i=1}^{n_b} H_{r-i} \times \left(P_{g_{i,t}} - P_{d_{i,t}}\right) \le \overline{f_r} r \in \Omega \& \forall t \in T \tag{23}$$

where H_{r-i} is the generalized distribution factor of line r with respect to bus i, $\overline{f_r}$ is the maximum transmission capacity for line r, and Ω is the set of transmission lines.

2.5. Probabilistic optimal power flow

A probabilistic OPF is a more appropriate approach when dealing with uncertainties of loads and wind power fluctuations, which process includes running the deterministic power flow continuously to account for the majority of possible system states. This chapter utilizes an approximate method called Hong's point estimate method (2 m + 1 scheme) to characterize uncertainties, which uses the first few front most statistical moments of stochastic variables to approximate the probability functions [23].

K is the number of concentration points that we use to represent the statistical information of the random input variable in our $K\times m$ scheme. A location $x_{i,k}$ and a weight $w_{i,k}(x_{i,k}, w_{i,k})$ represent the kth concentration of the random variable x_i. In order to relate the input and output variables to each other, we apply the non-linear function $F(x_1, x_2, \dots, x_i, \dots, x_m)$. The location of the kth value of variable x_i is determined by the following equation:

$$x_{i,k} = \mu_{x_i} + \xi_{i,k}\sigma_{x_i} \tag{24}$$

where μ_{x_i} is the mean for the input variable x_i, $\xi_{i,k}$ represents the standard location for the input variable x_i, and σ_{x_i} is the standard deviation for the input x_i. We assign a weighing factor $w_{i,k}$ to the current random output variable of the kth concentration. To determine $\xi_{i,k}$ and $w_{i,k}$, for the kth concentration of x_i, we use the following equations [23]:

$$\sum_{k=1}^{K} w_{i,k} = \frac{1}{m} \tag{25}$$

$$\sum_{k=1}^{K} w_{i,k}\left(\xi_{i,k}\right)^j = \lambda_{i,j} j = 1, \dots, 2K-1 \tag{26}$$

$\lambda_{i,j}$ in the equation above represents the jth standard central moment for the random variable x_i, and its probability density function f_{x_i} can be described as:

$$\lambda_{i,j} = \frac{M_j(x_i)}{\left(\sigma_{x_i}\right)^j} \tag{27}$$

The jth central moment of the random variable x_i is given by:

$$M_j(x_i) = \int_{-\infty}^{\infty} \left(x_i - \mu_{x_i}\right)^j f_{x_i} dx_i \tag{28}$$

Once we obtain every concentration $(x_{i,k}, w_{i,k})$, we use the nonlinear function F to calculate the vector of random output variables $Z(i,k)$ for each point $\left(\mu_{x_1}, \mu_{x_2}, \dots, x_{i,k}, \dots, \mu_{x_m}\right)$ as follows:

$$Z(i,k) = F\left(\mu_{x_1}, \mu_{x_2}, \dots, x_{i,k}, \dots, \mu_{x_m}\right) \tag{29}$$

By using the values from $Z(i,k)$, and the weighing factors, the jth moments of the random output variables can be approximated by:

$$E\left[Z^j\right] \cong \sum_{i=1}^{m} \sum_{k=1}^{K} w_{i,k}(Z(i,k))^j \tag{30}$$

We can extract the desired statistical information of our random output variable using a 2 m + 1 scheme by solving (25) for K = 3 and $\xi_{i,3} = 0$. The standard locations and weight produced by the equation are:

$$\xi_{i,k} = \frac{\lambda_{i,3}}{2} + (-1)^{3-k}\sqrt{\lambda_{i,4} - \frac{3}{4}\lambda_{i,3}{}^2}k = 1,2\ \xi_{i,3} = 0 \tag{31}$$

$$w_{i,k} = \frac{(-1)^{3-k}}{\xi_{i,k}(\xi_{i,1} - \xi_{i,2})}k = 1,2 \tag{32}$$

$$w_{i,3} = \frac{1}{m} - \frac{1}{\lambda_{i,4} - \lambda_{i,3}{}^2} \tag{33}$$

$\lambda_{i,3}$ and $\lambda_{i,4}$ are the skewness and kurtosis of x_i.

The scheme above sets up $\xi_{i,3} = 0$, which results in $x_{i,k} = \mu_{x_i}$ in (25), and yields m of the 3 m locations at the same point. By that done, it only requires one additional function evaluation for this particular location to complete 1 iteration of our probabilistic OPF. We update the corresponding weight to w_0 as follows:

$$w_0 = \sum_{i=1}^{m} w_{i,3} = 1 - \sum_{i=1}^{m} \frac{1}{\lambda_{i,4} - \lambda_{i,3}{}^2} \tag{34}$$

The deterministic DC-OPF is executed 2 m + 1 times in order to take all the random variables into account.

2.6. Reliability analysis

Reliability analysis provides an index to measure the degree of supply availability to meet the system demand. In the times when generated and stored energy is insufficient to supply the load, load is interrupted to maintain the power balance in the system. Load and generation variations as well as equipment failures are among the system uncertainties that could contribute to the load interruption in a power system. For wind turbines, the reliability model is a combination of a two-state model and power output model defined by (3). This combination is illustrated in **Figure 1** to provide the reliability model for wind generators.

The interrupted load in the system is equivalent to the amount of energy that is not supplied for each hour of the scheduling period, which can be described as:

$$ENS_t = \sum_{i=1}^{n_b} IL_{i,t} \qquad \forall t \in T \tag{35}$$

where ENS_t is the energy not supplied at hour t, and $IL_{i,t}$ is the interrupted load at bus i at hour t.

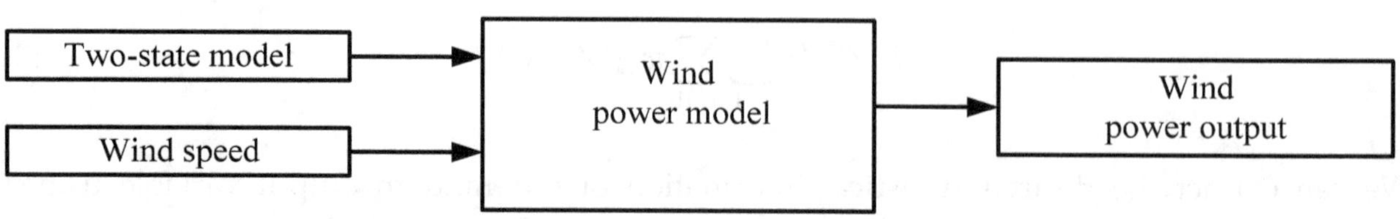

Figure 1. Reliability model for wind generator.

The interrupted load is defined as a random output variable whose first moment is calculated by (30), with j = 1. Expected energy not supplied (*EENS*) is then calculated for a one-year planning duration to provide a probabilistic index for our reliability analysis.

$$EENS = \sum_{c=1}^{C} \sum_{t=1}^{T} \sum_{i=1}^{n_b} E\left(IL_{i,t}^{c}\right).n_c \tag{36}$$

where n_c is the number of days within cluster c, C is the total number of clusters, c is the cluster number, and E is the average function.

We use the energy index of reliability (*EIR*) to estimate the reliability of the system, which can be calculated as follows:

$$EIR = 1 - \frac{EENS}{EE} \tag{37}$$

EE represents the expected energy demand of the system during the planning interval and is defined as:

$$EE = \sum_{c=1}^{C} \sum_{t=1}^{T} \sum_{i=1}^{n_b} P_{D_{i,t}}^{c}.n_c \tag{38}$$

2.7. Genetic algorithm optimization

We use a Genetic Algorithm (GA)-based optimization to install the energy storage with its optimal location and size. The GA begins by initially taking a set of randomly selected solutions, and then ranking the solutions based on their fitness values. We then perform recombination, crossover, selection, and mutation, to evolve the solution population. Once the satisfaction criterion is satisfied, we put the process into a halt. We assign a large penalty factor to the violated constraint to ensure satisfying constraints.

2.8. Proposed method

We model the storage system into our POPF as a load that stores excess, unconsumed energy generated by the system during off-peak periods. The storage system is modeled as a generator to release the stored energy to meet the peak load when sufficient transmission capacity is available. The location and scheduling of the storage systems are then optimized using GA. The optimized solution is the most cost efficient as it minimizes the total operation and interrupted-load costs for the span of the planning period. In order to optimally enhance the grid operability for wind integration, the storage technologies must possess an adequate capacity per the system's need. The fitness function that we use for the proposed method is the total weighted sum of the system's cost for each cluster over the planning period, as follows:

$$Fit.Function = \text{Min} \sum_{c=1}^{C} \sum_{t=1}^{T} (OC_t + ILC_t)_c n_c \tag{39}$$

where OC is the operation cost of the system, ILC is the interrupted-load cost of the system, and n_c represents the number of days within the cluster.

The proposed GA-based POPF can be described in the following steps:

1. Input wind speed, loads, and FOR data

2. Initialize the first population.

A. For t = 1, until t = T:

3. Initialize the first input variable by setting i = 1 and $E(Z) = 0$ & $E(Z^2) = 0$

B. For i = 1, until i = m:

4. Select input random variable x_i

5. Calculate $\xi_{i,k}$, $w_{i,k}$, $\lambda_{i,j}$

6. Initialize k=1.

C. For k = 1, until k = 3:

7. Calculate $x_{i,k}$

8a. If $G_{W_t} > L_t$, model the storage as a variable load with the following constraints:

$$0 \leq L_{S_t} \leq \min(S_{max} - (1 - d_s)S_{t-1}, P_{max})$$

8b. If $G_{W_t} < L_t$, model the storage as a generator with the following constraints:

$$0 \leq G_{S_t} \leq \min((1 - d_s)S_{t-1} - S_{min}, P_{max})$$

9. Run Deterministic OPF using $Z(i,k) = F\left(\mu_{x_1}, \mu_{x_2}, \ldots, x_{i,k}, \ldots, \mu_{x_m}\right)$

10. Calculate S for charging-discharging by using Eqs. (7 and 8)

11. Calculate OC_t, ILC_t, and $EENS_t$

12. Update raw moments using the following equations:

$$E(Z) = E(Z) + w_{i,k}Z(i,k); E(Z^2) = E(Z^2) + w_{i,k}[Z(i,k)]^2$$

13. If k = 3, go to step 14, if not, go to step **C** with $k = k + 1$.

14. If i = m, go to step 15, if not, go to step **B** with $i = i + 1$.

15. If t = T, go to step 16, if not, go to step **A** with $t = t + 1$.

16. Evaluate the fitness function and constraints

17. Generate children by using crossover and mutation

18a. If termination criteria are not met, produce next generation by selection and combination and go to step **2**.

18b. If termination criteria are met, calculate the statistical output information.

3. Case study

We evaluate our proposed method by applying it on the IEEE 24-bus system with the goal of solving for the optimal size and location for the storage units [10, 24]. In our simulations, we test our method for different situations by inputting different wind penetrations. To take into account

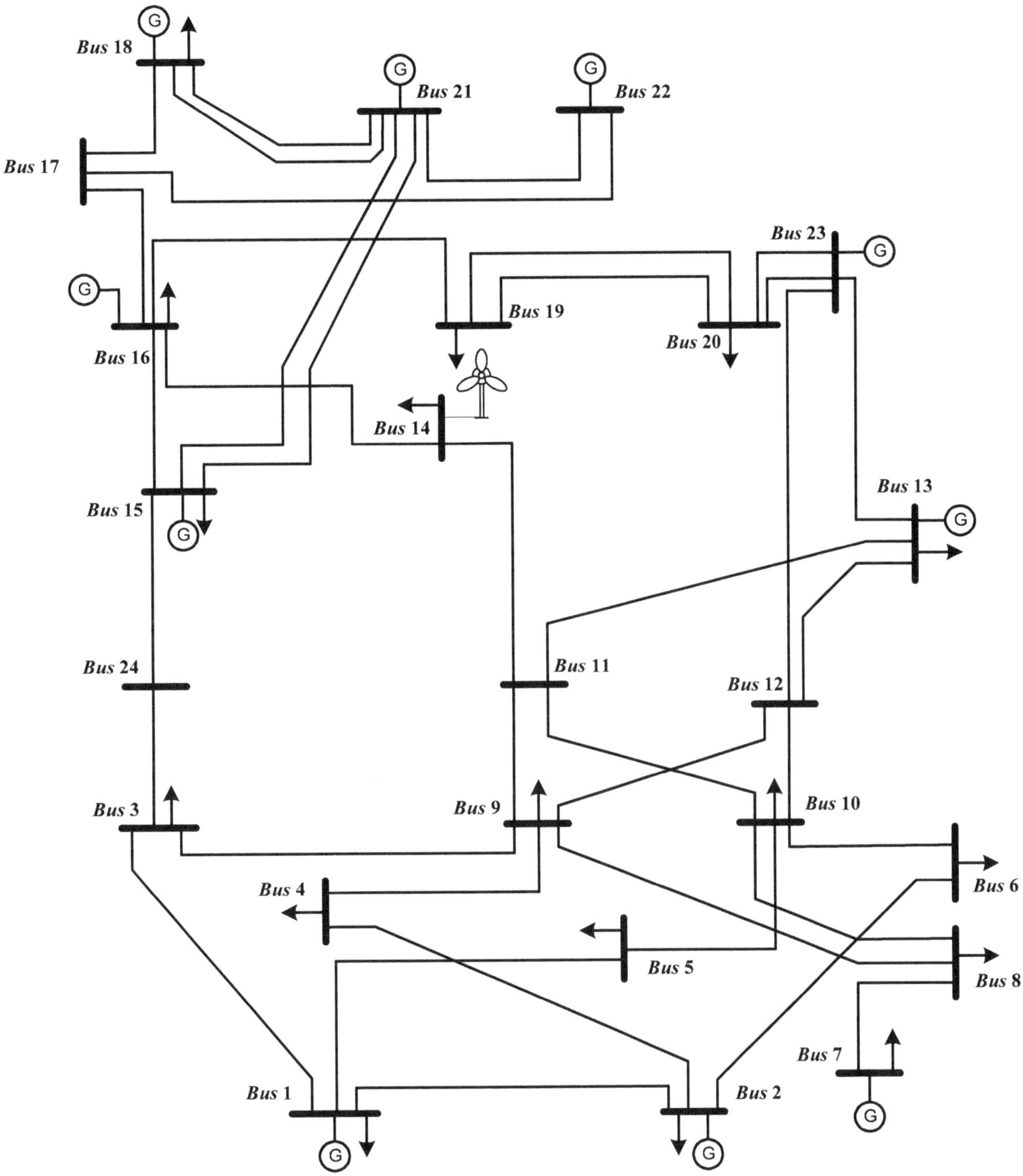

Figure 2. IEEE 24-bus system.

possible geological restrictions for CAES deployment in a real-world situation, we excluded busses 2, 7, 8, 11, and 17 from candidate locations on purpose. A wind farm is also predeterminedly installed at bus 14 in each of the case studies. A diagram of the IEEE 24-bus system is shown in **Figure 2**. **Tables 1–3** provide more information regarding the flow limitations, generators' cost functions, and IEAR values of the IEEE 24-bus system in use for the case study.

We define wind penetration (WP) as the ratio between the wind capacity installation and the system maximum load. We use real-world historical data obtained from the BPA for the system load [25] and from Mesonet (Ames Station) for the wind speed [26], to create a more realistic simulation environment. To calculate the cost of electric service reliability in the IEEE 24-bus system that we are going to run our demo on, we use the values of IEAR for our load busses [27]. The cost of the storage system, storage cost (SC), is equal to the sum of the cost of investment (*A*), and its cost of operation (OC_s) for the planning period. We can then calculate the cost of conventional generation (OC) by excluding the storage operation cost from the total operation cost.

The objective is to achieve the maximum possible reliability level. Our scenario's goal is to solve for the optimal placement and sizing for the storage system to meet the reliability objective. Our control strategy is to use the available wind energy to supply the load first, followed by utilizing the existing conventional generation capacity, and last, if necessary, to discharge power from the storage system to satisfy the load. The result of the simulations including the comparison with other conventional alternatives is shown in **Table 4**. Same reliability level is considered for both technologies to make a fair economic comparison. Our storage system, which enhances the reliability of wind integration, can be economically

From bus	To bus	Flow limit (MW)	From bus	To bus	Flow limit (MW)
1	2	175	11	13	500
1	3	175	11	14	500
1	5	175	12	13	500
2	4	175	12	23	500
2	6	175	13	23	500
3	9	175	14	16	500
3	24	400	15	16	500
4	9	175	15	21	500
5	10	175	15	24	500
6	10	175	16	17	500
7	8	175	16	19	500
8	9	175	17	18	500
8	10	175	17	22	500
9	11	400	18	21	500
9	12	400	19	20	500
10	11	400	20	23	500
10	12	400	21	22	500

Table 1. Transmission flow limitations.

Generator	Cost function coefficients		
	$a_i(\$/MW^2h)$	$b_i(\$/MWh)$	$c_i(\$)$
G_1	0.103	71.05	1313.6
G_2	0.108	71.04	1168.1
G_3	0.090	66.19	1078.8
G_4	0.091	67.26	969.8
G_5	0.078	71.60	958.2
G_6	0.078	71.60	958.2
G_7	0.100	73.90	471.6
G_8	0.090	73.90	471.6
G_9	0.098	69.70	445.4
G_{10}	0.101	66.51	702.7

Table 2. Coefficients of the cost function for the generators.

Bus No.	1	2	3	4	5	6	7	8
IEAR	6.20	4.89	5.30	5.62	6.11	5.50	5.41	5.40
Bus No.	9	10	11	12	13	14	15	16
IEAR	2.30	4.14	—	—	5.39	3.41	3.01	3.54
Bus No.	17	18	19	20	21	22	23	24
IEAR	—	3.75	2.29	3.64	—	—	—	—

Table 3. IEAR (*$/kWh*) values at each bus in the IEEE bus-system.

evaluated by comparing the sum of its associated costs with the total cost for the conventional alternative. The cost-benefit analysis from our simulation results shows the economic merits of the CAES, which can be found in **Table 4**.

4. Conclusions

Our case study concludes that energy storage technologies are more economic and technically sound options than fossil-fuelled generators to reliably and efficiently integrate intermittent renewable energy such as wind. The merits of energy storage application for reliability enhancement of renewable integration become even more highlighted when the emission costs associated with fossil-fuelled generators are included in the evaluation. This provides the subject of future studies.

	20% WP & 2500 MW PL*		30% WP & 3000 MW PL		40% WP & 3500 MW PL		50% WP & 4000 MW PL	
	CS**	CA***	CS	CA	CS	CA	CS	CA
Optimal Placement (*Bus #*)	14	—	14	—	23	—	23	—
S_{max} (*MWh*)	77.81	—	778.7	—	1524	—	1858	—
P_{max} (*MW*)	59.63	18.00	258.6	138.1	652.3	415.5	947.5	807.6
EIR (%)	99.80	99.80	99.28	99.28	99.03	99.03	99.19	99.19
Wind utilization (%)	94.49	86.42	88.69	78.58	80.26	70.99	72.06	62.22
OC (10^9\$)	0.588	0.603	0.620	0.661	0.656	0.711	0.696	0.774
GC (10^9\$)	—	0.002	—	0.012	—	0.037	—	0.077
ILC (10^9\$)	0.041	0.041	0.194	0.194	0.324	0.324	0.295	0.295
SC (10^9\$)	0.005	—	0.024	—	0.053	—	0.072	—

*Peak load.
**Centralized storage.
***Conventional alternative.

Table 4. Simulation results for different wind penetrations (WP).

Author details

Mahmoud Ghofrani* and Anthony Suherli

*Address all correspondence to: mrani@uw.edu

Electrical Engineering, Engineering and Mathematics Division, School of STEM, University of Washington Bothell, USA

References

[1] Kamalinia S, Shahidehpour M, Khodaei A. Security-constrained expansion planning of fast-response units for wind integration. Electric Power Systems Research. 2011;**810**:107-116

[2] Lee T. Optimal spinning reserve for a wind-thermal power system using EIPSO. IEEE Transactions on Power Systems. 2007;**22**(1):1612-1621

[3] Denholm P, Ela E, Kirby B, Milligan M, The Role of Energy Storage with Renewable Electricity Generation. Technical Report, NREL/TP-6A2-47187; 2010

[4] Zhao H, Wu Q, Hu S, Xu H, Rasmussen CN. Review of energy storage system for wind power integration support. Applied Energy. 2015;**137**:545-553

[5] Gyuk IP, Eckroad S. Energy Storage for Grid Connected Wind Generation Applications.

Washington, DC, EPRI-DOE Handbook Supplement, 1008703: U. S. Department of Energy; 2004

[6] Celli G, Mocci S, Pilo F, Loddo M. Optimal integration of energy storage in distribution networks. In: Proc. IEEE PowerTech Conf., Bucharest. 2009

[7] Bludszuweit H, Dominguez-Navarro JA. A probabilistic method for energy storage sizing based on wind power forecast uncertainty. IEEE Transactions on Power Systems. 2011;**26**(3):1651-1658

[8] Brekken TKA, Yokochi A, Jouanne AV, Yen ZZ, Hapke HM, Halamay DA. Optimal energy storage sizing and control for wind power applications. IEEE Transactions on Sustainable Energy. 2011;**2**(1):69-77

[9] Dutta S, Sharma R. Optimal storage sizing for integrating wind and load forecast uncertainties. In: Proc. IEEE PES Innovative Smart Grid Technologies (ISGT). 2012

[10] Ghofrani M, Arabali A, Etezadi-Amoli M, Fadala MS. Energy storage application for performance enhancement of wind integration. IEEE Transactions on Power Systems. 2013;**28**(4)

[11] Zou K, Agalgaonkar AP, Muttaqi KM, Perera S. Distribution system planning with incorporating DG reactive capability and system uncertainties. IEEE Transactions on Sustainable Energy. 2012;**3**(1):112-123

[12] Arabali A, Ghofrani M, Etezadi-Amoli M, Fadali MS, Baghzouz Y. Genetic- algorithm-based optimization approach for energy management. IEEE Transactions on Power Delivery. 2013;**28**(1):162-170

[13] Ketchen DJ, Shook CL. The application of cluster analysis in strategic management research: An analysis and critique. Strategic Management Journal. 1996;**17**:441-458

[14] Arabali A, Ghofarni M, Bassett JB, Moeini-Aghtaie M. Optimum sizing and siting of renewable energy based DG units in distribution systems. Optimization in Renewable Energy Systems: Recent Perspectives; 2017

[15] Nage GD. Analysis of wind speed distribution: Comparative study of Weibull to Rayleigh probability density function; a case of two sites in Ethiopia. American Journal of Modern Energy. 2016;**2**(3):10-16

[16] Sohoni V, Gupta SC, Nema RK. A critical review on wind turbine power curve modelling techniques and their applications in wind based energy systems. Journal of Energy. 2016;**2016**:1-18

[17] Billinton R, Allan RN. Reliability Evaluation of Power Systems. 2nd ed; 1984

[18] EPRI-DOE Handbook of Energy Storage for Transmission & Distribution Applications, EPRI, Palo Alto, CA, and the U.S. Washington, DC: Department of Energy; 2003

[19] Luo X, Wang J, Overview of Current Development on Compressed Air Energy Storage. EERA Technical Report; 2013

[20] Schoenung SM, Hassenzahl WV, Long- vs. short-term energy storage technologies analysis:

A life-cycle cost study: A study for the DOE energy storage systems program. Sandia National Laboratories, SAND2003-2783. 2003

[21] Das T, JD MC. Compressed Air Energy Storage. Ames, Iowa: Iowa State University; 2012

[22] A Report Prepared for Arizona Public Service Company, Study of Compressed Air Energy Storage with Grid and Photovoltaic Energy Generation. Arizona Research Institute for Solar Energy; 2010

[23] Hong HP. An efficient point estimate method for probabilistic analysis. Reliability Engineering and System Safety. 1998;**59**:261-267

[24] IEEE committee report, a reliability test system. IEEE Transactions on Power Apparatus and Systems. 1989;**4**(3):1238-1244

[25] Available from: http://transmission.bpa.gov/business/operations/wind/

[26] Available from: http://mesonet.agron.iastate.edu/agclimate/info.phtml

[27] Li Y. Bulk System Reliability Evaluation in a Deregulated Power Industry. A PhD Thesis Submitted to the Department of Electrical Engineering. University of Saskatchewan

3

Discretization of Random Fields Representing Material Properties and Distributed Loads in FORM Analysis

Ireneusz Czmoch

Abstract

The reliability analysis of more complicated structures usually deals with the finite element method (FEM) models. The random fields (material properties and loads) have to be represented by random variables assigned to random field elements. The adequate distribution functions and covariance matrices should be determined for a chosen set of random variables. This procedure is called discretization of a random field. The chapter presents the discretization of random field for material properties with the help of the spatial averaging method of one-dimensional homogeneous random field and midpoint method of discretization of random field. The second part of the chapter deals with the discretization of random fields representing distributed loads. In particular, the discretization of distributed load imposed on a Bernoulli beam is presented in detail. Numerical example demonstrates very good agreement of the reliability indices computed with the help of stochastic finite element method (SFEM) and first-order reliability method (FORM) analyses with the results obtained from analytical formulae.

Keywords: FORM, SFEM, discretization, random fields, reliability

1. Introduction

In general, the safety of a structure is analyzed in the space $\Omega_X = \{X \in R^n\}$ of basic random variables X. For a given failure mode or serviceability requirement, represented by the limit state surface $g(X) = 0$, the space Ω_X is divided into the safe subset, $\Omega^S = \{X \in R^n; g(X) > 0\}$, and the failure subset, $\Omega^F = \{X \in R^n; g(X) \le 0\}$. If all random variables are continuous with the multivariate joint probability density function $f_X(x)$, the failure probability is given by the integral

$$P_f = \int_{\Omega^F} f_x(\boldsymbol{x})d\boldsymbol{x} \tag{1}$$

The integral (Eq. (1)) can be evaluated exactly for a few cases with the most important one: the linear limit state surface and multidimensional normal distribution function of variables $\boldsymbol{X}$.

Development of reliability methods resulted in variety of powerful algorithms to estimate the probability of failure for complicated mechanical and statistical models of structures. The first-order reliability method (FORM) is the most popular approach applied in practice.

FORM algorithm starts with the nonlinear transformation. In general, non-normal random vector $\boldsymbol{X}$ is transformed into a standard normal (Gaussian) vector $\boldsymbol{Y}$ with zero mean and unit covariance matrix $\boldsymbol{C}_{YY} = \boldsymbol{I}$. The limit state surface $g(\boldsymbol{x}) = 0$ is mapped into a limit state surface $G(\boldsymbol{y}) = 0$. Next, the design point $\boldsymbol{y}^*$, that is, the point on the limit state surface with the minimum distance to the origin of the $\boldsymbol{Y}$ space, is determined by solving the nonlinear optimization problem with a nonlinear constraint $G(\boldsymbol{y}) = 0$

$$\beta = \min\sqrt{\boldsymbol{y}^T\boldsymbol{y}} \quad for \quad \boldsymbol{y} \quad on \quad G(\boldsymbol{y}) = 0 \tag{2}$$

The hyperplane tangential to the limit state surface at the point $\boldsymbol{y}^*$ is given by the formula

$$\beta - \boldsymbol{a}^T\boldsymbol{y} = 0 \tag{3}$$

where $\boldsymbol{a}$ is a unit outward normal vector to the hyperplane and β is the distance between the hyperplane and the origin (**Figure 1**). Since the random vector $\boldsymbol{Y} = \boldsymbol{Y}(\boldsymbol{X})$ has standard normal distribution, the first-order approximation of the failure probability is easily derived as follows

$$P_f \cong P\left[\beta - \boldsymbol{a}^T\boldsymbol{Y} \leq 0\right] = \Phi(-\beta) \tag{4}$$

The nonlinear constrained optimization problem (Eq. (2)) can be solved with many standard procedures as well as algorithms developed especially for this purpose, for example, algorithm for the case of independent, non-normal random variables [1], algorithm for problems with incomplete probability information [2].

All such solvers are iterative: for the assumed value of design point $\boldsymbol{x}^*_{(\mathrm{k})}$, the values of limit state function $g\left(\boldsymbol{x}^*_{(\mathrm{k})}\right)$ and its gradient $\nabla g\left(\boldsymbol{x}^*_{(\mathrm{k})}\right)$ are determined. Next, a new position of design point $\boldsymbol{x}^*_{(\mathrm{k+1})}$ is derived and the process continues until the convergence criteria are fulfilled. If the safety of mechanical problem is described by the limit state function with analytical form, then the gradient can be evaluated easily and one of the algorithms solving the optimization problem (Eq. (2)) can be applied directly. However, if the stochastic variability of material properties and loads is to be taken into account, SFEM approach must be applied.

In general, the limit state function $g(\boldsymbol{X}) = g(\boldsymbol{R}(\boldsymbol{X}), \boldsymbol{S}(\boldsymbol{X}))$ can be represented in terms of two vectors: resistance variables $\boldsymbol{R}$ and load effects $\boldsymbol{S}$. The elements of resistance variables vector $\boldsymbol{R}$

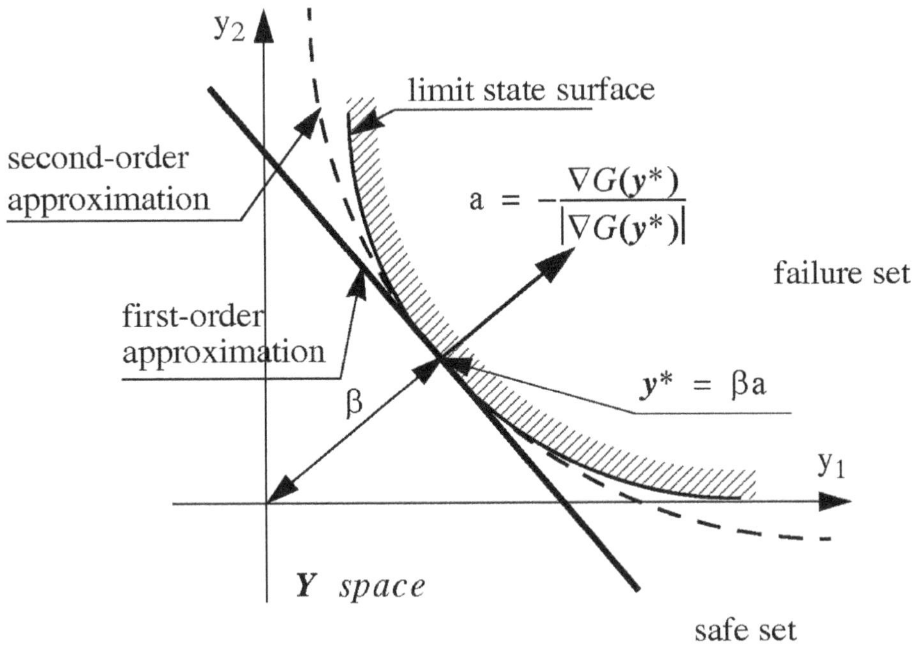

Figure 1. Definition of the FORM reliability index β.

(e.g., yield stress, allowable strain or allowable displacement), are prescribed to finite elements or nodes and can be treated as deterministic or random variables. In the latter case, the vector $\boldsymbol{R}$ corresponds to the part of the vector of basic random variables $\boldsymbol{X}$. The vector of load effects $\boldsymbol{S}$ (e.g., stresses, displacements and deformations) contains functions of basic random variables $\boldsymbol{X}$ such as material properties, geometrical quantities or loads. The relation $\boldsymbol{S} = \boldsymbol{S}(\boldsymbol{X})$ is called the mechanical transformation. In most practical cases, the load effects $\boldsymbol{S}$ have to be evaluated by using numerical algorithms, for example, FEM.

Two main problems are to be solved in order to apply FEM in FORM analysis:

- discretization of random fields of material properties and random fields of loads
- determination of the gradient of the limit state function $\nabla g(\boldsymbol{x}^*)$, when the load effect is defined by means of the implicit mechanical transformation $\boldsymbol{S} = \boldsymbol{S}(\boldsymbol{X})$

The solution of the second problem is presented in many papers and books [3].

2. Probabilistic description of random fields

2.1. Basics definitions

The spatial probabilistic variability of physical quantities such as Young's modulus, thickness of a plate and intensity of a distributed load can be described by means of random fields, $w(\boldsymbol{z})$, where $\boldsymbol{z}$ is the vector of space coordinates. One-dimensional random fields can be defined for beams, bars and columns, two-dimensional random fields for plates or shells, and three-dimensional random fields for bodies.

For any specific location z, random field $w(z)$ is a random variable with the cumulative distribution function

$$F_w(w(z)) = P[w(z) \leq w] \tag{5}$$

which is called the first-order distribution of the random field $w(z)$.

The $m-th$ order distribution, that is, the joint cumulative distribution function of the random vector $\boldsymbol{w} = [w(z_1), \ldots, w(z_m)]^T$, is defined as follows:

$$F_{w\ldots w}(\boldsymbol{w}) = P[w(z_1) \leq w_1, \ldots, w(z_m) \leq w_m] \tag{6}$$

The first- and second-order probability density functions of the random field $w(z)$ are defined accordingly

$$\begin{aligned} f_w(w(z)) &= \frac{d}{dw} F_w(w(z)) \\ f_{ww}(w(z_1), w(z_2)) &= \frac{\partial^2}{\partial w(z_1) \partial w(z_2)} F_{ww}(w(z_1), w(z_2)) \end{aligned} \tag{7}$$

Following the well-known definition [4] with the help of the first-order probability density function (Eq. (6)) and the second-order probability density function (Eq. (7)), the second-order representation of the random field is defined by using the following functions: the mean value function $\mu_w(z)$, the variance function $\sigma_w^2(z)$, the covariance function, $C_w(z_1, z_2)$ and the correlation function $\rho_w(z_1, z_2)$.

A random field $w(z)$ is called *strict-sense homogeneous*, if its statistics are invariant to the translation of the origin and in particular, the $n-th$ order density function has the property

$$f_{w\ldots w}(w_1(z_1), \ldots, w_n(z_n)) = f_{w\ldots w}(w_1(z_1 + z), \ldots, w_n(z_n + z)) \tag{8}$$

for any separation vector z.

A random field $w(z)$ is called *wide-sense homogeneous* or *second-order homogeneous* if its mean value and variance are constant,

$$\mu_w(z) = \mu_w \qquad \sigma_w^2(z) = \sigma_w^2 \tag{9}$$

and its covariance function as well as correlation function depends only on the separation vector z,

$$C_w(z, 0) = C_w(z_1, z_1 + z) = C_w(z) \quad for \quad any \quad z \tag{10}$$

A random field that is homogeneous in time is referred to as *stationary process*.

2.2. Ensemble average versus spatial average of random field

In order to estimate the statistical parameters of a random field, the sample (realization) must be collected in separate experiments. If the sample size is sufficiently large, the estimators of statistical parameters can be computed at each point of the random field domain. For example, at the location z_1, the estimator of mean value and the estimator of variance are equal to ensemble averages

$$\widehat{\mu}_w(z_1) = \frac{1}{k}\sum_{i=1}^{k} w_{(i)}(z_1) \tag{11}$$

$$\widehat{\sigma}_w^2(z_1) = \frac{1}{k-1}\sum_{i=1}^{k} \left[w_{(i)}(z_1) - \widehat{\mu}_w(z_1)\right]^2 \tag{12}$$

where k is the number of realizations (also called as the sample size) and $w_{(i)}(z)$ is the $i-th$ measurement of a random field.

Ensemble averages usually depend on the location vector. However, if the limit of the ensemble averages are invariant with respect to location

$$\widehat{\mu}_w(z_1) = \lim_{k\to\infty}\frac{1}{k}\sum_{i=1}^{k} w_{(i)}(z_1) = \lim_{k\to\infty}\frac{1}{k}\sum_{i=1}^{k} w_{(i)}(z_2) = \mu_w \tag{13}$$

then the random field can be considered as homogeneous, in strict- or wide-sense, which depends on the order of probability function invariant to location vector.

On the other hand, the spatial averages over the domain can be computed for every realization (measurement) of random field. For example, the average taken along with any single realization of *a one-dimensional random field* is equal to

$$\mu_w(i, L) = \frac{1}{L}\int_0^L w_{(i)}(z)dz \tag{14}$$

and it usually depends on the character of field and the length of averaging interval L.

A homogeneous random field is called ergodic, if all statistical information can be obtained from one realization of the random field. This means that ensemble averages are invariant with respect to the location vector and the spatial averages are equal to the ensemble averages. Thus, in case of a homogeneous one-dimensional random field, the ensemble and spatial averages are equal in the limit

$$\lim_{k\to\infty}\frac{1}{k}\sum_{i=1}^{k} w_{(i)}(z) = \lim_{L\to\infty}\frac{1}{L}\int_0^L w_{(i)}(z)dz \tag{15}$$

In general, it is usually difficult to prove that a random field is homogeneous, and it is even more difficult to prove that a random field is ergodic. Great number of samples over a sufficiently large domain should be collected. These conditions are rarely fulfilled. Thus, the homogeneity and ergodicity is usually assumed. Most of the concepts and methods developed in the reliability analysis are based on these assumptions.

2.3. One-dimensional homogeneous random field

A one-dimensional homogeneous random field is often used in the reliability analysis of linear elements such as beams, bars and frames. All the above-mentioned definitions are valid in this case.

The variance reduction function $\gamma_w(L)$, which has been presented in detail in [5, 7], describes the correlation of the moving average $w_L(z)$ of one-dimensional homogeneous random field $w(z)$

$$w_L(z) = \frac{1}{L}\int_{z-L/2}^{z+L/2} w(l)dl \tag{16}$$

where L denotes the length of the averaging segment.

The mean value function and the variance function of the random field $w_L(z)$ are easy to determine

$$E[w_L(z)] = E[w(z)] = \mu_w \tag{17}$$

$$Var[w_L(z)] = \sigma_L^2 = \gamma_w(L)\sigma_w^2 \tag{18}$$

where $\gamma_w(L)$ is the variance reduction function, and μ_w and σ_w^2 are mean value and variance value of the one-dimensional homogeneous random field $w(z)$. The variance reduction function $\gamma_w(L)$ demonstrates how fast the point variance σ_w^2 is reduced under local averaging. This dimensionless function has the following properties:

$$\gamma_w(L) = \gamma_w(-L) \geq 0 \qquad \gamma_w(0) = 1 \tag{19}$$

and is related to the correlation function $\rho_w(z)$ of the one-dimensional homogeneous random field $w(z)$ by the integral

$$\gamma_w(L) = \frac{1}{L^2}\int_0^L\int_0^L \rho_w(l_1 - l_2)dl_1 dl_2 = \frac{2}{L}\int_0^L \left[1 - \frac{l}{L}\right]\rho_w(l)dl \tag{20}$$

Another useful scalar measure of the correlation is the scale of fluctuation θ_w defined by the limit value of the variance reduction function

$$\theta_w = \lim_{L\to\infty} L\gamma_w(L) \tag{21}$$

It can be proved that the scale of fluctuation θ_w is related to the correlation function $\rho_w(z)$

$$\theta_w = 2\int_0^\infty \rho_w(l)dl \tag{22}$$

The variance reduction function and the scale of fluctuation are especially useful in the discretization procedure of the homogeneous random field.

Table 1 presents four correlation models. It should be noticed that the rectangular and triangular models are not proper correlation functions for the homogeneous random field, since they do not fulfill the basic condition of weak-homogeneity. However, they are quite often assumed, mostly as visualization tools. Triangular model demonstrates the meaning of the scale of fluctuation in a simple way, that is, correlation between values of random field at points separated by greater distance than the scale of fluctuation is equal to zero. The rectangular model constitutes the upper limit for variance reduction functions. The simple form of the exponential correlation function makes analytical computation of many integrals possible. On the other hand, similarity between the squared exponential model and the triangular model allows the simple physical interpretation of the scale of fluctuation, that is, the correlation functions are equal to zero for a separation interval greater than the scale of fluctuation.

A special case of random field is the Gaussian random field, in which the random variables $w(z_1), \ldots, w(z_n)$ for any points $z_1, \ldots, z_n$ are jointly normal distributed. This random field is completely determined by two functions such as the mean value function and the covariance function. The *n-th* order probability density function has the joint normal density.

Figure 2 presents correlation functions and **Figure 3** presents variance reduction functions for the correlation models described in **Table 1**.

Model	Correlation function	Variance function
Rectangular	$\rho(z) = \begin{cases} 1 & \lvert z\rvert \le \frac{\theta}{2} \\ 0 & \lvert z\rvert > \frac{\theta}{2} \end{cases}$	$\gamma(L) = \begin{cases} 1 & L \le \frac{\theta}{2} \\ \left(\frac{\theta}{L}\right)\left[1 - \frac{\theta}{4L}\right] & L > \frac{\theta}{2} \end{cases}$
Triangular	$\rho(z) = \begin{cases} 1 - \frac{\lvert z\rvert}{\theta} & \lvert z\rvert \le \theta \\ 0 & \lvert z\rvert > \theta \end{cases}$	$\gamma(L) = \begin{cases} 1 - \frac{L}{3\theta} & L \le \theta \\ \left(\frac{\theta}{L}\right)\left[1 - \frac{\theta}{3L}\right] & L > \theta \end{cases}$
Exponential	$\rho(z) = exp\left(-2\frac{\lvert z\rvert}{\theta}\right)$	$\gamma(L) = \frac{1}{2}\left(\frac{\theta}{L}\right)^2\left[\frac{2L}{\theta} - 1 + exp\left(-2\frac{L}{\theta}\right)\right]$
Squared exponential	$\rho(z) = exp\left(-\pi\left(\frac{z}{\theta}\right)^2\right)$	$\gamma(L) = \frac{2\theta}{L}\left[\Phi\left(\frac{L\sqrt{2\pi}}{\theta}\right) - 0,5\right] + A$ $A = \frac{\theta^2}{\pi L^2}\left[exp\left(-\pi\left(\frac{L}{\theta}\right)^2\right) - 1\right]$

Note: θ is the scale of fluctuation; L is the separation (averaging) distance; and $\Phi(z)$ is the Laplace function.

Table 1. Description of four correlation models.

3. Discretization of random fields representing material properties

A vast amount of papers deal with the problem how to develop the accurate and numerically efficient discretization methods for random fields of material properties.

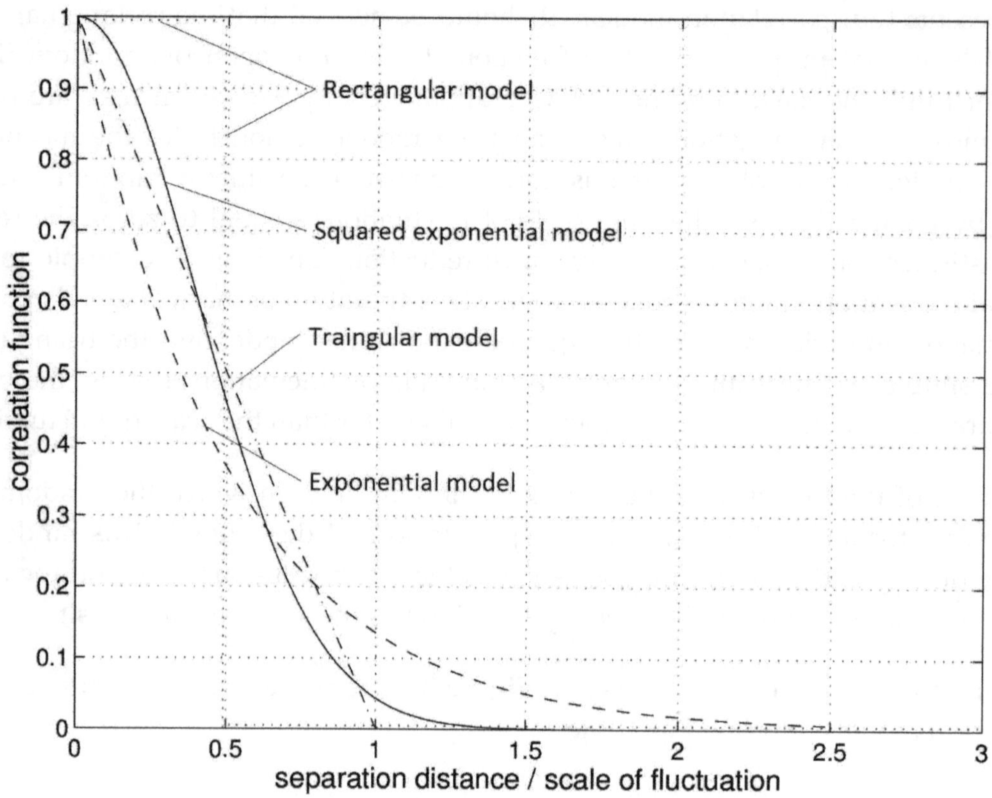

Figure 2. Four correlation functions for correlation models defined in **Table 1.**

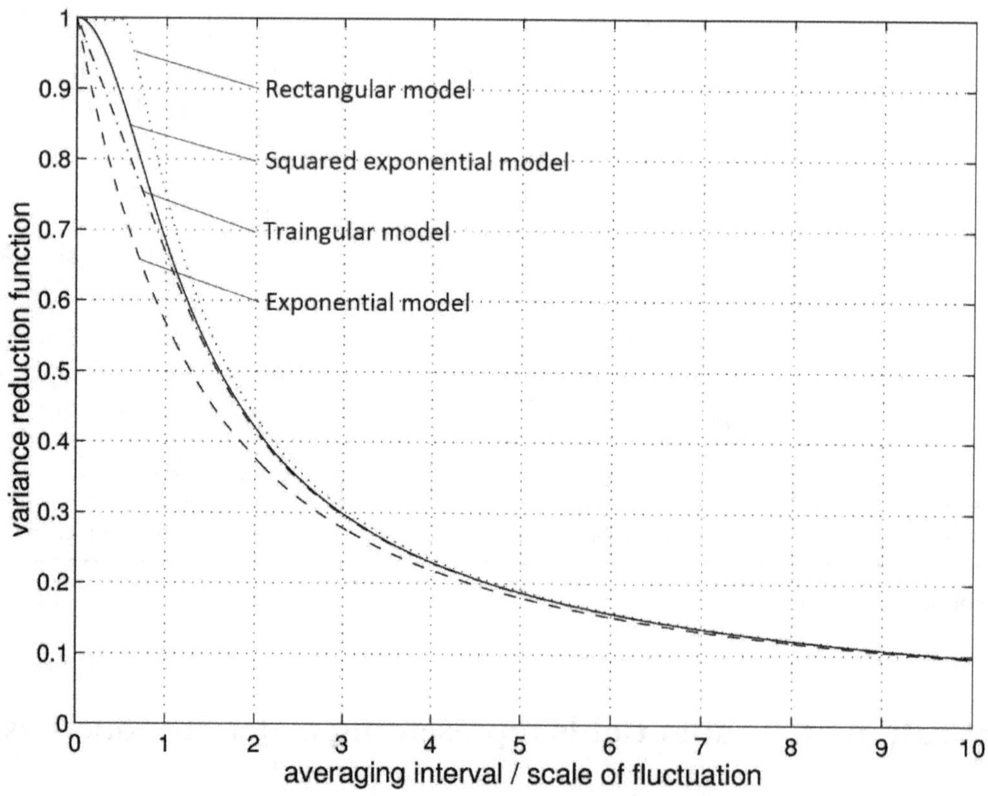

Figure 3. Four variance reduction functions for correlation models defined in **Table 1**.

The variability of random field is usually more accurately represented, if the number of random field elements or the number of series components is increased. However, greater number of random variables leads to longer computation time for realistic problems. Therefore, it has been an important issue to find out the optimal size of random field elements with respect to the scale of fluctuation, that is the scalar correlation measure. The accuracy of different methods is discussed by Zeldin and Spanos [9].

The reliability analysis of more complicated structures usually deals with FEM models. The random fields (material properties and loads) have to be represented by random variables assigned to random field elements. The adequate distribution functions and covariance matrices should be determined for a chosen set of random variables. This procedure is called discretization of a random field.

Two groups of methods for discretization of material random fields can be distinguished:

1. Random field elements

The value of a material property for any finite element is represented by a single random variable, constant within a random field element. Mean value, standard deviation and covariance as well as distribution function can be assigned to those random variables according to different procedures:

- the spatial averaging method [10]
- the midpoint method [8, 9, 11]
- the interpolation method [13]

2. Random series

The random field is described in terms of series of deterministic functions and random coefficients. Two examples of this approach are as follows:

- series composed of deterministic shape functions and random variables [16]
- the Karhunen-Loeve orthogonal expansion [15, 17, 20]

Two discretization methods, namely the spatial averaging method and the midpoint method, are presented in detail.

3.1. The spatial averaging method of one-dimensional homogeneous random field

The spatial averaging method has been developed by Vanmarcke [5]. We consider a one-dimensional homogeneous random field $w(z)$ that represents the spatial random variability of a material property, for example, modulus of elasticity along beam. In general, the domain of the random field can be divided into finite elements of lengths L_i. The material property within the $i-th$ element is represented by a random variable which is assumed to be equal to the spatial average over the $i-th$ finite element

$$W_i = \frac{1}{L_i}\int_0^{L_i} w(z)dz \tag{23}$$

The mean value of a random variable W_i is equal to the mean value of the random field $w(z)$

$$E[W_i] = \frac{1}{L_i}\int_0^{L_i} E[w(z)]dz = \mu_w \tag{24}$$

and the variance of random variable W_i is expressed in terms of the variance function $\gamma_w(L)$ of the random field $w(z)$

$$Var[W_i] = \gamma_w(L_i)\sigma_w^2 \tag{25}$$

The formula for the covariance between two random variables W_i and W_j related to the *i-th* and *j-th* random elements is more complicated

$$Cov\left[W_i, W_j\right] = \frac{\sigma_w^2}{2L_iL_j}\sum_{k=0}^{3}(-1)^k L_k^2\gamma(L_k) \tag{26}$$

where the distances L_k are defined in **Figure 4**.

Eqs. (25 and 26) can be generalized for a random field defined in two- or three-dimensional spaces [5]. Eq. (26) depends on the variance reduction function $\gamma_w(L)$, which expresses a relation between the variance of the spatial average and the size of the averaging interval L.

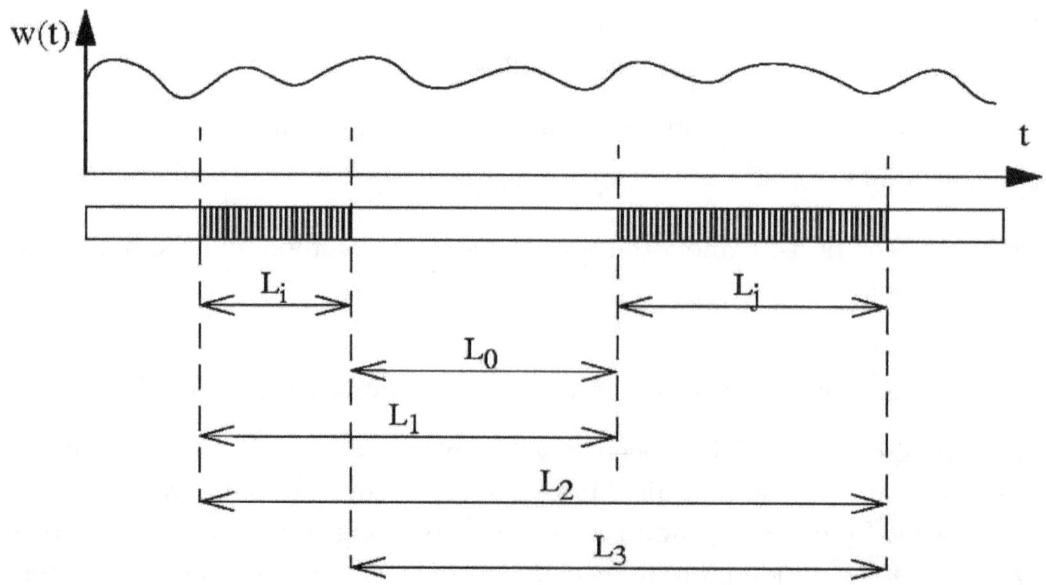

Figure 4. Definition of intervals used in the calculation of covariance between the spatial averages related to two random field elements.

Since full information about the variability of the random field is seldom available, Vanmarcke [5–6] suggested using in the practical analysis the approximation of the variance reduction function by its asymptotic form

$$\gamma_w(L) = \begin{cases} 1 & L \leq \theta_w \\ \dfrac{\theta_w}{L} & L > \theta_w \end{cases} \tag{27}$$

where θ_w is the scale of fluctuation.

FORM analysis demands the knowledge about distribution functions of basic random variables. The spatial averaging method results in the normal random variables for the Gaussian random field $w(z)$, since the integration is a linear operation. However, for non-Gaussian random field, it is difficult to derive the distribution function of a random variable W_i defined by Eq. (23).

Der Kiureghian [18] has suggested a heuristic model for the distribution of random variable W_i, which is based on the concept of the weighting the random field by the shape function, which results in the weighted variance reduction function. **Figure 5** shows that the weighted variance function has much smaller values than the original variance reduction function. If the averaging interval $L_i = L_{long}$ is assumed many times longer than the scale of fluctuation θ_w, the distribution of random variable W_i tends to have the normal distribution according to the central limit theorem. Then, the variance reduction function can be approximated as follows:

$$\gamma_w\left(L_{long}\right) \approx \frac{\theta_w}{L_{long}} \approx \frac{1}{n} \quad for \quad L_{long} \gg \theta_w \tag{28}$$

where the parameter $n \gg 1$ should be determined by calculations or judgment.

According to **Figure 3**, which shows the variance functions of four models, as well as taking into account **Figure 5**, $n \approx 10$ could be assumed, which means that for the element length which is 10 times longer than the scale of fluctuation, the normal distribution can be assigned to the random variable W_i. For shorter elements, non-normal distribution should be considered. On the other hand, for very short element length, the distribution of random variable W_i is close to the first-order distribution of random field. Taking into account both limits the approximate density distribution function of W_i has been proposed by Der Kiureghian [18]

$$f_{W_i}(w_i) = \alpha \frac{\sqrt{n}}{\sigma_w} \varphi\left(\frac{w_i - \mu_w}{\sigma_w/\sqrt{n}}\right) + (1-\alpha) f_{w(z)}(w_i) \tag{29}$$

where $\varphi(u)$ is the standard normal probability density function and $f_{w(z)}(w)$ is the first-order probability density function of random field $w(z)$.

The weight parameter $0 \leq \alpha \leq 1$ can be determined for the current length L_i of random field element by requiring that the variance of random variable W_i with the approximate density

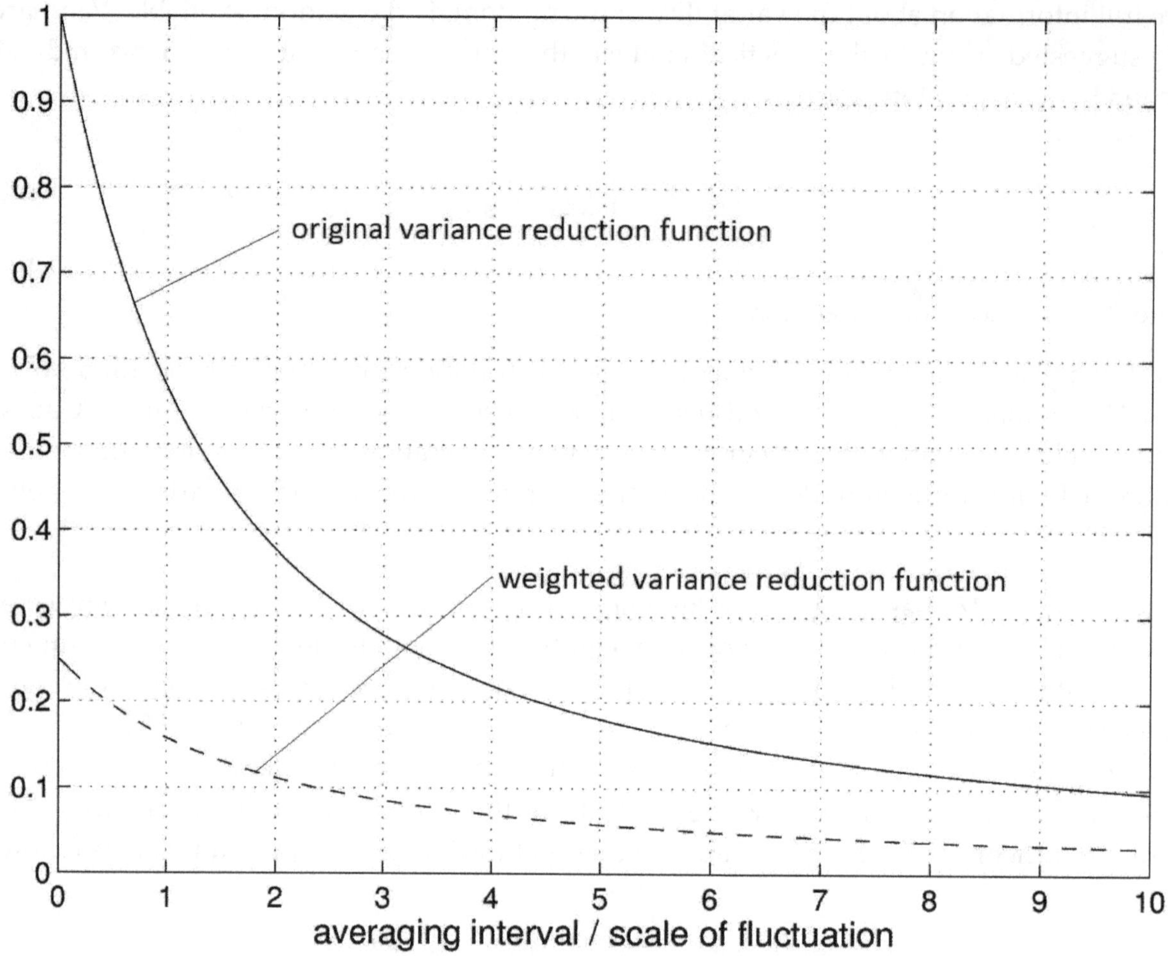

Figure 5. The original and weighted variance reduction functions for the random field with exponential correlation function.

function ((Eq. (29)) is equal to the variance of the spatial average of random field over the length L_i, that is, $\gamma_w(L_i)\sigma_w^2$. It can be shown that for the element of length L_i

$$\alpha = \frac{n}{n-1}(1 - \gamma_w(L_i)) \tag{30}$$

3.2. Midpoint method of discretization of random field

The midpoint method [3] corresponds to the interpolation method [13] with constant interpolation function and is suitable for discretization of non-Gaussian random fields. In general, a nonhomogeneous random field $w(z)$ is discretized with the help of a set of random variables defined as follows:

$$W_i = w\left(z_c^{(i)}\right) \tag{31}$$

where $z_c^{(i)}$ determines the position of the centroid of the i-th element.

The mean value of random variable W_i and the covariance between random variables W_i and W_j are given below, correspondingly

$$E[W_i] = E\left[w\left(z_c^{(i)}\right)\right] \qquad Cov\left[W_i, W_j\right] = C_w\left(z_c^{(i)}, z_c^{(j)}\right) \tag{32}$$

where $C_w(z_1, z_2)$ is the covariance function of nonhomogeneous random field.

In the midpoint discretization method, the probability distribution of the random variable W_i is equivalent to the first-order distribution of the random field $w(z)$

$$F_{W_i}(w_i) = P[W_i \le w_i] = P\left[w\left(z_c^{(i)}\right) \le w_i\right] = F_w\left(w_i\left(z_c^{(i)}\right)\right) \tag{33}$$

Thus, in case of a homogeneous random field, the probability distribution function does not depend on the location of the centroid of the i-th element, $F_{W_i}(w_i) = F_w(w_i)$.

3.3. Selection of the optimal size for random field mesh representing variability of a material property

Both discretization procedures described in the previous section are based on assumption that property (e.g., modulus of elasticity) is constant within a finite element, see [19].

In deterministic FEM, the variability of material properties is modeled by means of sufficient number of finite elements. The structural finite element size is chosen with respect to the gradient of the stress field.

In the same way, the variability of the random field is usually more accurately represented if many random field elements are used in the analysis. However, finer random field mesh increases the number of random variables, which leads to longer computation times for the reliability analysis. Therefore, it has been an important issue to find out the optimal size of the random field elements.

The scale of fluctuation of the random field has been shown to be a very important measure of the correlation, since it governs the optimal size of a random field mesh. Der Kiureghian and Ke [12] have shown that a sufficiently accurate value of the reliability index is obtained if the random field element size is between one-half and one-quarter of the scale of fluctuation for the midpoint method with the exponential correlation function. Hisada and Nakagiri [11] have presented similar results.

If random field elements shorter than one-quarter of the scale of fluctuation are chosen, then a singular correlation matrix can be obtained, indicating linear dependency of the random variables. Then, the nonhomogeneous linear transformation to the set of uncorrelated, normalized basic random variables must be proceeded by an extra transformation, which decreases the dimension of the random variable space. However, this extra transformation is not unique. Improper choice of the transformation can lead to numerical difficulties in the iteration procedure for determining the reliability index. Therefore, too small random field elements should be avoided. Liu and Liu [19] have derived a simple rule of thumb regarding the selection of an appropriate random field mesh: a coarse mesh should be assumed in an area where the gradient of the limit state function with respect to the random variable representing random field is small and a finer random field mesh in an area with large gradient.

However, in many case a random element size equal to the scale of fluctuation may be considered as adequate with respect to the reliability analysis accuracy.

4. Discretization of random fields representing distributed loads

Discretization of random field loads has not been a subject of many studies. The finite element modeling introduces the well-known procedure for representing distributed forces $\boldsymbol{p}(\boldsymbol{s})$ by a set of equivalent nodal forces. The random field of distributed loads has to be discretized according to the structural finite element mesh. Thus, if the distributed loads are random, all equivalent nodal forces become random variables. This approach seems obvious and it can be applied directly to study the response variability of stochastic engineering problems [14]. However, if discretized random field load is a part of FORM calculations, then both the discretization procedure for random field loads as well as FORM/FEM analysis should be modified in order to get accurate results.

The general approach for the discretization procedure of the distributed body forces $\boldsymbol{p}(\boldsymbol{s})$ is presented below [20]. A similar algorithm can be applied for other types of distributed loads (surface forces and initial stresses). For the purpose of FORM analysis, it is convenient to assume that the nodal forces for *i-th* element are random variables

$$\boldsymbol{Q}_p^{(i)} = \int_V \mathbf{N}^{(i)^T} \boldsymbol{p}(\boldsymbol{s}) dV \tag{34}$$

where $\mathbf{N}^{(i)}$ is the displacement interpolation matrix for the *i-th* element in the local coordinate system. Thus, the mean value vector of the load vector $\boldsymbol{Q}_p^{(i)}$ is equal to

$$E\left[\boldsymbol{Q}_p^{(i)}\right] = \int_V \mathbf{N}^{(i)^T} E[\boldsymbol{p}(\boldsymbol{s})] dV \tag{35}$$

The covariance matrix of the vectors of equivalent forces $\boldsymbol{Q}_p^{(i)}$ *and* $\boldsymbol{Q}_p^{(j)}$ corresponding to the *i-th* and *j-th* finite elements has the form

$$Cov\left[\boldsymbol{Q}_p^{(i)}, \boldsymbol{Q}_p^{(j)\mathrm{T}}\right] = \int_{V^{(i)}} \int_{V^{(j)}} \mathbf{N}^{(i)^T} \mathbf{C}_{pp} \mathbf{N}^{(j)} dV^{(i)} dV^{(j)} \tag{36}$$

where the matrix $\boldsymbol{C}_{pp}$ contains the cross-covariance functions between different components of the vector random field.

In general, a load effect can be an internal force, stress or strain at any point of structure which does not coincide with a nodal point (or Gaussian integration point). Thus, the load effect

$S^{(i)}(s)$ at a point s of *i-th* element can be represented as a sum of a general solution, $S_u^{(i)}(s)$ and a particular solution, $S_p^{(i)}(s)$. The general solution $S_u^{(i)}(s)$ is the load effect as a function of geometry, material properties and equivalent nodal forces applied at all nodal points. Whereas the particular solution $S_f^{(i)}(s)$ is the load effect at the point s of the *i-th* element due to the distributed body forces $f^{(i)}(s)$ and reactions $Q_p^{(i)}$ at the *i-th* element.

In the FORM analysis, the vector of basic random variables $\boldsymbol{X}$ also contains nodal equivalent forces $Q_p^{(i)}$ (where the parameter "i" runs over all finite elements). Thus, in the search for the most likely failure point, the current values of equivalent nodal forces $Q_p^{(i)}$ have to be determined at each iteration step of FORM algorithm. Those equivalent nodal forces, valid at a specific step of FORM iteration, correspond to unknown functions of distributed body forces $p(s)$. In order to determine the particular solution $S_p^{(i)}(s)$, the function of the distributed body forces $p^{(i)}(s)$ within the $i-th$ finite element must be known. One way to solve this problem is to assume that the distributed body forces can be approximated with the help of shape functions.

$$\boldsymbol{p}^{(i)}(s) = \boldsymbol{N}^{(i)^T}\boldsymbol{b} \tag{37}$$

The matrix $\boldsymbol{b}$ can be determined from the condition that the equivalent nodal force $Q_p^{(i)}$ at the *i-th* finite element due to body forces $p^{(i)}(s)$ should be equal to the calculated equivalent nodal forces in the FORM algorithm. The vector $\boldsymbol{b}_k$, which is the *k-th* column of the matrix $\boldsymbol{b}$ and corresponds to the *k-th* component of the vector $p^{(i)}(s)$, is determined by solving the system of linear equations

$$\left(\int_V \boldsymbol{N}_k^{(i)^T}\boldsymbol{N}_k^{(i)} dV\right)\boldsymbol{b}_k = \boldsymbol{Q}_p^{(i)} \tag{38}$$

where $N_k^{(i)}$ is the *k-th* row of the matrix $N^{(i)}$.

In this way, the function of the distributed body forces $p^{(i)}(s)$ as well as the particular solution $S_p^{(i)}(s)$ is determined as functions of nodal equivalent forces $Q_p^{(i)}$.

For the distributed loads represented by the Gaussian random field, the components of vector $Q_p^{(i)}$, which are determined by means of a linear transformation (Eq. (34)), have the multi-dimensional normal distribution. For the non-Gaussian random field, the probability distribution function $F_{Q_p^{(i)}}(r)$ of the vector of nodal equivalent forces $Q_p^{(i)}$ cannot be determined easily. The first possible choice is to assume the normal distribution on the basis of the central limit theorem. Another approximate solution has been developed [20] on the basis of the approach presented by Der Kiureghian [18].

5. Discretization of transverse distributed load for a Bernoulli-Euler beam

In case of Bernoulli-Euler beam, four shape functions are applicable

$$N_1(s) = 1 - 3\left(\tfrac{s}{L}\right)^2 + 2\left(\tfrac{s}{L}\right)^3 \quad N_2(s) = s\left(1 - \tfrac{s}{L}\right)^2$$
$$N_3(s) = \left(\tfrac{s}{L}\right)^2\left(3 - 2\frac{s}{L}\right) \quad N_4(s) = \frac{s^2}{L}\left(\frac{s}{L} - 1\right) \tag{39}$$

The nodal forces equivalent to the transverse distributed load q(s) are defined by the integrals

$$Q_j = \int_0^L N_j(s)q(s)ds \quad j = 1, \ldots, 4 \tag{40}$$

The distributed load q(s) is assumed to be a homogeneous random field, with constant mean μ_q value, constant variance σ_q^2 and the covariance function for the correlation function ρ_q

$$\mathrm{Cov}[q(s), q(t)] = \sigma_q^2\rho_q(|t - s|) \tag{41}$$

Thus, the mean value of the nodal force Q_i is just equal to

$$E[Q_i] = \mu_q \int_0^L N_i(s)ds \tag{42}$$

and the covariance between nodal force $Q_i^{(m)}$ at the $m - th$ finite element and the nodal force $Q_i^{(n)}$ at the $n - th$ finite element is defined by the double integral

$$\mathrm{Cov}\left[Q_i^{(m)}, Q_j^{(n)}\right] = \sigma_q^2 \int_{s_{1m}}^{s_{2m}} \int_{s_{1n}}^{s_{2n}} N_i(s)N_j(t)\rho_q(|t - s|)dsdt \tag{43}$$

where s_{1m}, s_{2m} and s_{1n}, s_{2n} are the coordinates of the two ends of the finite elements, defined in a common coordinate system.

If a beam is divided into *N* finite elements, then the random distributed load q(s) is modeled by *4N* random variables.

A typical limit state function for a beam can be defined at the $m - th$ finite element

$$g^{(m)}(s) = R_b^{(m)}(s) - M^{(m)}(s) \tag{44}$$

where $R_b^{(m)}(s)$ is the bending resistance at cross-section s (usually assumed as a basic random variable) and $M^{(m)}(s)$ is the bending moment due to external loads at cross-section s, which is a random function depending on the other basic random variables, for example, random nodal equivalent forces $Q_1^{(1)}, Q_2^{(1)}, Q_3^{(1)}, Q_4^{(1)}, \ldots Q_1^{(N)}, Q_2^{(N)}, Q_3^{(N)}, Q_4^{(N)}$.

The bending moment $M^{(m)}(s)$ can be represented as a sum

$$M^{(m)}(s) = M_u^{(m)}(s) + M_q^{(m)}(s) \tag{45}$$

The general solution $M_u^{(m)}(s)$ depends on the nodal displacements, which are the functions of all nodal forces $Q_1^{(1)}, Q_2^{(1)}, Q_3^{(1)}, Q_4^{(1)}, \ldots Q_1^{(N)}, Q_2^{(N)}, Q_3^{(N)}, Q_4^{(N)}$ imposed to FEM model. The particular solution $M_q^{(m)}(s)$ is the bending moment within the $m-th$ element due to the distributed load q(s) and reactions $-Q_1^{(m)}, -Q_2^{(m)}, -Q_3^{(m)}, -Q_4^{(m)}$ at the $m-th$ element,

$$M_q^{(m)}(s) = Q_2^{(m)} - Q_1^{(m)} s - \int_{s_{1m}}^{s} q(t)(s-t)dt \tag{46}$$

At the $k-th$ iteration step of FORM algorithm, the function of the distributed load is unknown for the corresponding nodal forces. Therefore, the function of distributed load imposed on the $m-th$ finite element is assumed as a linear combination of the shape functions

$$q(s) \cong q^{(m)}(s) = \sum_{i=1}^{4} b_i N_i(s) \tag{47}$$

and the unknown coefficients b_i, which have to be determined by requiring that the equivalent nodal forces (Eq. (40)) for the function $q^{(m)}(s)$ defined by relation (Eq. (47)) should be equal to the current equivalent nodal forces. The coefficients b_i as well as the distributed load $q^{(m)}$ can be obtained as functions of the current equivalent nodal forces $Q_1^{(m)}, Q_2^{(m)}, Q_3^{(m)}, Q_4^{(m)}$ at the $m-th$ finite element, by solving the system of linear equations

$$\sum_{i=1}^{4}\left[\int_{s_{1m}}^{s_{2m}} N_i(s)N_j(s)ds\right] b = Q_j^{(m)} \quad j = 1, \ldots, 4 \tag{48}$$

Finally, the bending moment $M_q^{(m)}(s)$ is determined as a function of equivalent nodal forces

$$M_q^{(m)}(s) = m_1(s)Q_1^{(m)} + m_2(s)Q_2^{(m)} + m_3(s)Q_3^{(m)} + m_4(s)Q_4^{(m)} \tag{49}$$

where

$$\begin{aligned}
m_1(s) &= -s\left(1 - 8\frac{s}{L} + 20\left(\frac{s}{L}\right)^2 - 20\left(\frac{s}{L}\right)^3 + 7\left(\frac{s}{L}\right)^4\right) \\
m_2(s) &= 1 - 60\left(\frac{s}{L}\right)^2 + 200\left(\frac{s}{L}\right)^3 - 225\left(\frac{s}{L}\right)^4 + 84\left(\frac{s}{L}\right)^5 \\
m_3(s) &= -s\left(2\frac{s}{L} - 10\left(\frac{s}{L}\right)^2 + 15\left(\frac{s}{L}\right)^3 - 7\left(\frac{s}{L}\right)^4\right) \\
m_4(s) &= -30\left(\frac{s}{L}\right)^2 + 140\left(\frac{s}{L}\right)^3 - 195\left(\frac{s}{L}\right)^4 + 84\left(\frac{s}{L}\right)^5
\end{aligned} \tag{50}$$

In order to determine the reliability index β_{FORM}, the gradient of the limit state function (Eq. (44)) has to be calculated. The partial derivatives of function (Eq. (44)) with respect to the nodal forces are given below

$$\frac{\partial g^{(m)}}{\partial Q_i^{(n)}} = -\sum_{j=1}^{4} \frac{\partial M_u^{(m)}}{\partial u_j^{(m)}} \frac{\partial u_j^{(m)}}{\partial Q_i^{(n)}} \quad if \quad m \neq n \quad i = 1, \ldots, 4 \tag{51}$$

$$\frac{\partial g^{(m)}}{\partial Q_i^{(m)}} = -\sum_{j=1}^{4} \frac{\partial M_u^{(m)}}{\partial u_j^{(m)}} \frac{\partial u_j^{(m)}}{\partial Q_i^{(m)}} - m_i(s) \qquad i = 1, \ldots, 4 \tag{52}$$

where $u_j^{(m)}$ is the nodal displacements in the local coordinate system of the $m-th$ element and the derivatives $\frac{\partial u_j^{(m)}}{\partial Q_i^{(n)}}$ are computed with the help of the SFEM algorithm (Liu Der Kiureghian, 1991).

5.1. Example of discretization of random distributed load for a simply supported beam

The deterministic simply supported beam of length L is subjected to the transverse homogeneous random load q(s) with mean value μ_q and variance σ_q^2. Assuming the exponential correlation function with the scale of fluctuation θ_q,

$$\rho(\tau) = \exp\left(-2\frac{|\tau|}{\theta_q}\right) \tag{53}$$

the mean value and the variance of the bending moment function can be derived analytically:

$$E[M(s)] = \mu_q \frac{s(L-s)}{2} \qquad Var[M(s)] = \sigma_q^2 \left[A\frac{\theta_q^4}{8} + B\frac{L\theta_q^3}{4} + C\frac{L^3\theta_q}{3} \right] \tag{54}$$

where

$$A = \left(1 - \frac{s}{L}\right)\left[\frac{s}{L} exp\left(-2\frac{L}{\theta_q}\right) - exp\left(-2\frac{s}{\theta_q}\right)\right] + \frac{s}{L}\left[\frac{s}{L} - exp\left(-2\frac{L-s}{\theta_q}\right) - 1\right] + 1$$

$$B = -\left(1 - \frac{s}{L}\right)\frac{s}{L} \qquad C = \left(1 - \frac{s}{L}\right)^2\left(\frac{s}{L}\right)^2$$

We consider the linear limit state function $g(s) = R_b - M(s)$ where R_b is the deterministic bending moment capacity, constant along the beam.

If the random field q(s) is Gaussian, then the FORM reliability index is equivalent to the Cornell reliability index

$$\beta_{FORM}(s) = \beta_C(s) = \frac{R_b - \mathrm{E}[M(s)]}{\sqrt{\mathrm{Var}[M(s)]}} \tag{55}$$

On the other hand, the FORM reliability indices have been computed for the finite element model of a simply supported beam. The distributed load random field q(s) has been discretized according to the procedure described earlier.

The calculations have been carried out for the following data:

$\mu_q = 1000$ N/m, $\sigma_q = 200$ N/m, $R_b = 5000$ Nm, L = 6 m and 5 cross-sections: s = [1.02; 1.5; 2.04; 2.52; 3] (m). Three finite element sizes have been considered: $L_e = 0.5; 1; 3$ (m). The scale of fluctuation has been assumed as: $\theta_q = 0.5; 1; 4; 100$ (m)

The results of the FORM analysis presented in **Table 2** are in very good agreement with the reliability indices computed according to analytical formulae (Eq. (53–55)). Moreover, the

Finite element length	Position of cross-section (m)				
(m)	1.02	1.50	2.04	2.52	3.00
Scale of fluctuation = 0.5					
Analytical	14.7350	7.3059	3.6040	2.1244	1.6821
0.5	14.7347	7.3055	3.6038	2.1243	1.6820
1.0	14.7345	7.3055	3.6038	2.1243	1.6820
3.0	14.7511	7.3166	3.6047	2.1247	1.6819
Scale of fluctuation = 1.0					
Analytical	10.7242	5.3037	2.6112	1.5378	1.2172
0.5	10.7240	5.3036	2.6112	1.5377	1.2172
1.0	10.7239	5.3036	2.6112	1.5377	1.2172
3.0	10.7314	5.3092	2.6116	1.5378	1.2172
Scale of fluctuation = 4.0					
Analytical	6.5774	3.2516	1.5997	0.9415	0.7451
0.5	6.5773	3.2517	1.5997	0.9415	0.7451
1.0	6.5773	3.2516	1.5997	0.9415	0.7451
3.0	6.5781	3.2523	1.5997	0.9415	0.7451
Scale of fluctuation = 100.0					
Analytical	4.9153	2.4423	1.2062	0.7114	0.5633
0.5	4.9153	2.4423	1.2062	0.7114	0.5633
1.0	4.9153	2.4423	1.2062	0.7114	0.5633
3.0	4.9153	2.4423	1.2062	0.7114	0.5633

Table 2. Reliability indices computed according to analytical formula and FORM algorithm.

reliability indices computed with the help of the described discretization procedure are insensitive to the scale of fluctuation and the finite element size.

6. Conclusions

The reliability analysis of more complicated structures usually deals with the FEM models. The random fields (material properties and loads) have to be represented by random variables assigned to random field elements. The adequate distribution functions and covariance matrices should be determined for a chosen set of random variables. This procedure is called discretization of a random field.

The chapter presents the discretization of random field for material properties with the help of the spatial averaging method of one-dimensional homogeneous random field, and midpoint method of discretization of random field.

The second part of the chapter deals with the discretization of random fields representing distributed loads. The discretization of distributed load imposed on a Bernoulli beam is presented in detail. An example shows that the presented procedure for discretizing random fields representing distributed loads is very efficient, that is, the reliability indices computed with the help of SFEM and FORM analysis are in very good agreement with the results of analytical calculations.

Author details

Ireneusz Czmoch

Address all correspondence to: i.czmoch@il.pw.edu.pl

Department of Structural Mechanics and Computer Aided Engineering, Faculty of Civil Engineering, Warsaw University of Technology, Warsaw, Poland

References

[1] Rackwitz R, Fiessler B. Structural reliability under combined load sequences. Computers and Structures. 1978;**9**:489-494

[2] Der Kiureghian A, Liu PL. Structural Reliability Under Incomplete Probability Information. Journal of Engineering Mechanics Division, ASCE. Jan. 1986;**112**(1):85-104

[3] Liu P-L, Der Kiureghian A. Finite element reliability of geometrically nonlinear uncertain structures. Journal of Engineering Mechanics Division, ASCE. 1991;**117**(8):1806-1825

[4] Bury KV. Statistical Methods in Applied Science. John Wiley & Sons; 1975

[5] Vanmarcke EH. Random Fields: Analysis and Synthesis. Cambridge: Mass. and London Massachusetts Institute of Technology Press; 1983

[6] Vanmarcke EH. Random field modeling of the void phase of soils. Georisk. March 2007;**1**(1): 57-68

[7] Knabe W, Przewłócki J, Różyński G. Spatial averages for linear elements for two-parameter random field. Probabilistic Engineering Mechanic. 1998;**1**:147-167

[8] Huang S, Mahadaevan S, Rebba R. Collocation-based stochastic finite element analysis for random field problems. Probabilistic Engineering Mechanics. 2007;**22**:194-205

[9] Zeldin BA, Spanos PD. On Random Field Discretization in Stochastic Finite Elements. Journal of Applied Mechanics. 1998;**65**:320-327

[10] Vanmarcke EH, Grigoriu M. Stochastic finite element analysis of simple beams. Journal of Engineering Mechanics, ASCE. 1983;**109**(5):1203-1214

[11] Hisada T, Nakagiri S. Role of the Stochastic Finite Element Method in Structural Safety and Reliability. 4th Int. Conf. on Structural Safety and reliability; Kobe, Japan. 1985. pp. 385-394

[12] Der Kiureghian A, Ke B-J. The stochastic finite element method in structural realiability. Probabilistic Engineering Mechanics. 1988;**3**(2):83-91

[13] Liu WK, Belytschko T, Mani A. Random field finite elements. International Journal for Numerical Methods in Engineering. 1986;**23**:1831-1845

[14] Yang LF, Yu B, Ju JW. System reliability analysis of spatial variance frames based on random field and stochastic elastic modulus reduction method. Acta Mech. 2012;**223**:109-124

[15] Allaix DL, Carbone VI. An efficient coupling of FORM and Karhunen-Loeve series expansion. Engineering with Computers. 2016;**32**:1-13

[16] Lawrence M. Basic random variables in finite element analysis. International Journal for Numerical Methods in Engineering. 1987;**24**:1849-1863.

[17] Spanos PD, Ghanem R. Stochastic finite element expansion for random media. Journal of Engineering Mechanics, ASCE. 1989;**115**(5):1035-1053

[18] Der Kiureghian A. Multivariate distribution models for structural reliability. In: Wittmann FH, editor. Transaction of the 9-th conference on structural mechanics in reactor technology. Vol. M. 17–21 Aug 1987. Lausanne. pp. 373-379

[19] Liu P-L, Liu K-G. Selection of random field mesh in finite element reliability analysis. Journal of Engineering Mechanics, ASCE. 1993;**119**(4):667-680

[20] Czmoch I. Influence of Structural Timber Variability on Reliability and Damage Tolerance of Timber Beams. Lulea University of Technology; 1998

4

Modeling Strategies to Improve the Dependability of Cloud Infrastructures

Erica Teixeira Gomes de Sousa and
Fernando Antonio Aires Lins

Abstract

Cloud computing presents some challenges that need to be overcome, such as planning infrastructures that maintain availability when failure events and repair activities occur. Cloud infrastructure planning that addresses the dependability aspects is an essential activity because it ensures business continuity and client satisfaction. Redundancy mechanisms cold standby, warm standby and hot standby can be allocated to components of the cloud infrastructure to maintain the availability levels agreed in service level agreement (SLAs). Mathematical formalisms based on state space such as stochastic Petri nets and based on combinatorial as reliability block diagrams can be adopted to evaluate the dependability of cloud infrastructures considering the allocation of different redundancy mechanisms to its components. This chapter shows the adoption of the mathematical formalisms stochastic Petri nets and reliability block diagrams to dependability evaluation of cloud infrastructures with different redundancy mechanisms.

Keywords: dependability evaluation, state space models, non-state space models, redundancy mechanisms, maintenance policies

1. Introduction

Ensuring the availability levels required by the different services hosted in the private cloud is a great challenge. The occurrence of defects in these services can cause the degradation of their response times and the interruption of service of a request due to unavailability of the required resource. The interruption of these services can be caused by the occurrence of failure events in the hardware, software, power system, cooling system and private cloud network. When the occurrence of defects is constant, users give less preference to hiring service providers due to reduced availability, reliability and performance of these services [1].

The dependability assessment can minimize the occurrence of faults and failure events [2] in the private cloud and promote the levels of availability and reliability defined in the SLAs, avoiding the payment of contractual fines. One option to ensure the availability of services offered in the private cloud is to assign redundant equipment to its components. Redundant devices allow service reestablishment, minimizing the effects of failure events. The major problem with this assignment is the estimation of the number of redundant equipment and the choice of the type of redundancy that must be considered to guarantee the quality of the service offered. The estimation of the type and number of redundant equipment should also consider the cost of the quantitative of each type of redundancy mechanism attributed to the components of the cloud computing [3, 4].

2. Basic concepts

The dependability evaluation denotes the ability of a system to deliver a reliably service. Dependability measures are reliability, availability, maintainability, performability, safety, testability, confidentiality, and integrity [2].

Dependability evaluation is related to the study of the effect of errors, defects and failures in the system, since these have a negative impact on the dependability attributes. A fault is defined as the failure of a component, subsystem or system that interacts with the system in question [5]. An error is defined as a state that can lead to a failure. A defect represents the deviation from the correct operation of a system. A summary of the main measures of dependability is shown below.

The reliability of a system is the probability (P) that this system performs its function satisfactorily, without the occurrence of defects, for a certain period of time (T). Reliability is represented by Eq. (1), where T is a random variable that represents the time for occurrence of defects in the system [3, 4].

$$R(t) = P\{T > t\}, t \geq 0 \tag{1}$$

The probability of the occurrence of defects up to a time t, is represented by Eq. (2), where T is a random variable that represents the time for system failures [3, 4].

$$F(t) = 1 - R(t) = P\{T \leq t\} \tag{2}$$

Eq. (3) represents the reliability, considering the density function F(t) of the time for occurrence of failures (T) in the system [3, 4, 6].

$$R(t) = P\{T > t\} = \int_t^{\infty} F(t)dt \tag{3}$$

The Mean Time to Failure (MTTF) is the average time for defects to occur in the system. When this average time follows the exponential distribution with parameter λ, the MTTF is represented by Eq. (4) [3, 4, 6].

$$\mathrm{MTTF} = \int_0^{\infty} R(t)dt = \int_0^{\infty} e^{(-\lambda)t} = \frac{1}{\lambda} \tag{4}$$

The failures can be classified in relation to the time, according to the mechanism that originated them. The behavior of the failure rate can be represented graphically through the bathtub curve, which presents three distinct phases: infant mortality (1), useful life (2) and aging (3). **Figure 1** shows the variation of the failure rate of hardware components as a function of time [7].

During the infant mortality phase (1), a reduction in the failure rate occurs. Failures during this period are due to equipment manufacturing defects. In order to shorten this period, manufacturers submit the equipment to a process called burn-in, where they are exposed to high operating temperatures. In the useful stage (2), the failures occur randomly. Equipment reliability values provided by manufacturers apply to this period. The service life of the equipment is not normally a constant. It depends on the level of stress in which the equipment is subjected during that period. During the aging phase (3), an increase in the failure rate occurs.

In high availability environments, one must be sure that the infant mortality phase has passed. In some cases, it is necessary to leave the equipment running in a test environment during this time. At the same time, care must be taken to have the equipment replaced before entering the aging phase.

The availability of a system is the probability that this system is operational for a certain period of time, or has been restored after a defect has occurred. Uptime is the period of time in which the system is operational, downtime is the period of time when the system is not operational due to a defect or repair activity occurring, and uptime + downtime is the time period of observation of the system. Eq. (5) represents the availability of a system [3, 4, 6].

$$A = \frac{\text{uptime}}{\text{uptime} + \text{downtime}} \tag{5}$$

Computational systems and applications require different levels of availability and therefore can be classified according to these levels. U.S. Federal Aviation Administration's National Airspace System's Reliability Handbook classifies computer systems and applications according to their criticality levels [1]. These computational systems and applications can be

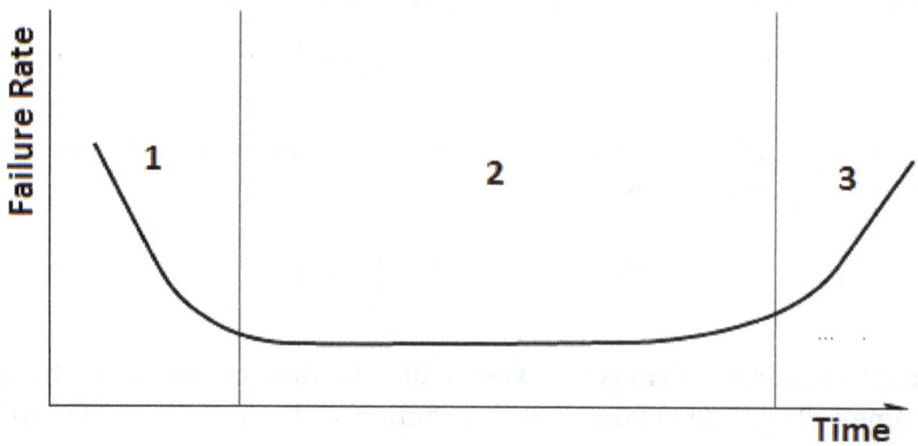

Figure 1. Bathtub curve.

considered critical critics when the required availability is 99.99999%, critical when the required availability is 99.999%, essential when the required availability is 99.9% and routine when the required availability is 99% [1].

The maintainability is the probability that a system can be repaired in a given period of time (TR). The maintainability is described by Eq. (6), where TR denotes the repair time. This equation represents maintainability, since the repair time TR has a density function G(t) [3, 4, 6].

$$V(t) = P\{TR \leq tr\} = \int_0^{tr} G(t)dt \tag{6}$$

The Mean Time to Repair (MTTR) is the average time to repair the system. When the time distribution function of repair is represented by an exponential distribution with parameter μ, the MTTR is represented by Eq. (7) [3, 4, 6].

$$MTTR = \int_0^{\infty} 1 - G(tr)dt = \int_0^{\infty} 1 - e^{(\mu)tr} = \frac{1}{\mu} \tag{7}$$

Mean Time Between Failures (MTBF) is the mean time between system defects, represented by Eq. (8) [3, 4, 6].

$$MTBF = MTTR + MTTF \tag{8}$$

Performability describes the degradation of system performance caused by the occurrence of defects [3, 6].

3. Redundancy mechanisms

The redundancy mechanisms provide greater availability and reliability to the system during the occurrence of failure events due to the maintenance of components operating in parallel, that is, a redundant system has a secondary component that will be available when the primary component fails. Thus, redundancy mechanisms are designed to avoid single points of failure and therefore provide high availability and disaster recovery if necessary [1, 8].

The redundancy mechanisms can be classified as active-active and active-standby. Active-active redundancy mechanisms are employed when the primary and secondary components share the workload of the system. When any of these components fails, the other component will be responsible for servicing the system users' requests. These redundancy mechanisms can be classified as N + K, where K secondary components identical to N primary components are required for system workload sharing. In the N + 1, configuration, a secondary component identical to the primary N components is required for sharing the system workload. In the N + 2, configuration, two secondary components identical to the N primary components are required for system workload sharing [1].

The active-standby redundancy mechanisms are employed when the primary components meet the requests of the system users and the secondary components are on hold. When the primary components fail, the secondary components will be responsible for servicing the system users' requests. The active-standby redundancy mechanisms can be classified as hot standby, cold standby and warm standby [1].

In the hot standby redundancy mechanism, redundant modules that are in standby function in synchronization with the operating module, without their computation being considered in the system, and in case the occurrence of a failure event is detected, it is ready to make operational immediately [1, 4].

In the cold standby redundancy mechanism, the redundant modules are turned off and only when a failure event occurs will they be activated after a time interval. In the cold standby redundancy mechanism, inactive modules that are de-energized, by hypothesis, do not fail, whereas the active module has a constant failure rate λ.

In the warm standby redundancy mechanism, the redundant modules that are in standby function in sync with the operating module, without their computation being considered in the system. If a fault event is detected, the redundant module is ready to become operational after a time interval. Systems with standby sparing of cold standby sparing and warm standby sparing need more time for recovery compared to hot standby sparing, but systems with cold standby sparing and warm standby sparing have the advantage of lower power consumption and no wear standby systems [1, 4].

4. Modeling techniques

The models adopted for dependability evaluation can be classified as combinatorial and state space. The combinatorial models capture the conditions that cause failures in the system or allow its operation when considering the structural relationships of its components. The best known combinatorial models are Reliability Block Diagram (RBD) and Fault Tree (FT) [9, 10]. State space-based models represent the behavior of the system (occurrence of failures and repair activities) through its states and the occurrence of events. These models allow the representation of dependency relations between the components of the systems. The most widely used state space-based models are Markov Chains (MC) and Stochastic Petri Net (SPN) [9–12].

The SPN models provide great flexibility in the representation of aspects of dependability. However, these models suffer from problems related to the size of the state space for computational systems with large number of components [11, 12]. RBD models are simple, easy to understand, and their solution methods have been extensively studied. These models can represent the components of cloud computing that do not have a dependency relation to allow an efficient representation, avoiding growth problems too much of the space of states [9, 10].

5. Reliability block diagram

Reliability block diagram (RBD) is one of the most used techniques for reliability analysis of systems [5].

The RBD allows the calculation of availability and reliability by means of closed formulas, since it is a combinational model. These closed formulas make the calculation of the result faster than the simulation, for example [6].

In a reliability block diagram, components are represented with blocks combined with other blocks (i.e., components) in series, parallel or combinations of those structures. A diagram that has components connected in series requires each component to be running for the system to be operational. A diagram that has components connected in parallel requires that only one component is working for the system to be operational [13]. Thus, the system is described as a set of interconnected functional blocks to represent the effect of availability and reliability of each block on the availability and reliability of the system [14].

The availability and reliability of two blocks connected in series is obtained through Eq. (9) [6].

$$P_s = \prod_{i=1}^{n} P_i(t) \tag{9}$$

where:

Pi(t) describes the reliability Ri(t), the instantaneous availability Ai(t) e a and the steady state availability Ai of the block Bi.

The availability and reliability of two blocks connected in parallel is obtained through Eq. (10) [6].

$$P_p = 1 - \prod_{i=1}^{n} (1 - P_i(t)) \tag{10}$$

where Pi(t) describes the reliability Ri(t), the instantaneous availability Ai(t) e a and the steady state availability Ai of the block Bi.

Figure 2 shows the connection of the blocks in series and **Figure 3** shows the connection of the blocks in parallel.

The reliability block diagram is mainly used in modular systems consisting of many independent modules, where each can be easily represented by a block.

Figure 2. Reliability block diagram in series.

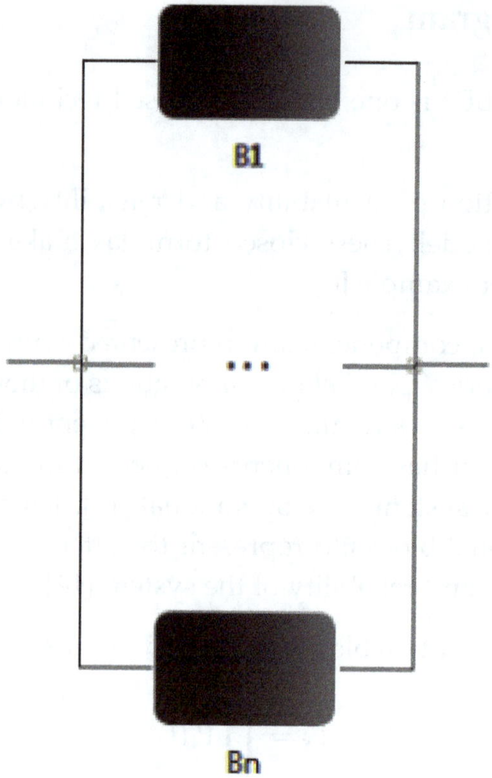

Figure 3. Reliability block diagram in parallel.

6. Petri nets

The concept of Petri nets was introduced by Carl Adam Petri in 1962 with the presentation of his doctoral thesis "Kommunikation mit Automaten" (Communication with Automata) [15] at the Faculty of Mathematics and Physics of Darmstadt University in Germany. Petri nets are graphical and mathematical tools used for formal description of systems characterized by properties of concurrency, parallelism, synchronization, distribution, asynchronism, and non-determinism [15].

The applicability of Petri nets as a tool for systems study is important because it allows for mathematical representation, analysis of models and also for providing useful information about the structure and dynamic behavior of the modeled systems. The applications of Petri nets can occur in many areas (systems of manufacture, development and testing of software, administrative systems, among others) [16].

The Petri nets presents some characteristics that are: the dynamic representation of the modeling system with the desired level of detail; The graphical and formal description that allows to obtain information on the behavior of the modeled system through its behavioral and structural properties; The representation of synchronism, asynchronism, competition, resource sharing, among other behaviors; And the wide applicability and documentation.

Petri nets are formed by places (1), transitions (2), arcs (3) and marking (4). The places correspond to state variables and the transitions, actions or events performed by the system. The performance of an action is associated with some preconditions, that is, there is a relation between the places and the transitions that allows or not the accomplishment of a certain action. After performing a certain action, some places will have their information changed, that is, the action will create a post condition. The arcs represent the flow of the marking through the Petri net, and the tokens represent the state in which the system is at a given moment. Graphically, places are represented by ellipses or circles, transitions, by rectangles, arcs, by arrows and marking, by means of dots (**Figure 4**) [16].

The two elements, place and transition, are interconnected by directed arcs as shown in **Figure 5**. The arcs that interconnect places to the transitions (Place → Transition) correspond to the relationship between the true conditions (precondition), which enable the execution Of the shares. The arcs that interconnect transitions to places (Transition → Place) represent the relationship between actions and conditions that become true with the execution of actions (post condition) [16].

The formal mathematical representation of a model in Petri net (Petri net—PN) is the quintuple PN = P, T, F, W, M0 [15], where:

6.1. Properties of Petri nets

The study of the properties of Petri nets allows the analysis of the modeling system. Property types can be divided into two categories: the initial marking-dependent properties, named behavioral properties, and the non-marking properties, named structural properties [15, 16].

6.1.1. Behavioral properties

The behavioral properties are those that depend only on the initial marking of the Petri net. The properties covered are reachability, limitation, safeness, liveness and coverage.

Reachability indicates the possibility that a given marking can be reached by firing a finite number of transitions from an initial marking. Given a Petri net marked RM = (R,M0), the triggering of a transition t0 alters the marking of the Petri net. An M′ label is accessible from M0 if there is a sequence of transitions which, triggered, lead to the M′ label. That is, if the marking M0 enables the transition t0, by triggering this transition, the marking M1 is reached. The marking M1 enables t1 which, upon being triggered, reaches the marking M2 and so on until the marking M′ is obtained.

Figure 4. Elements of Petri net.

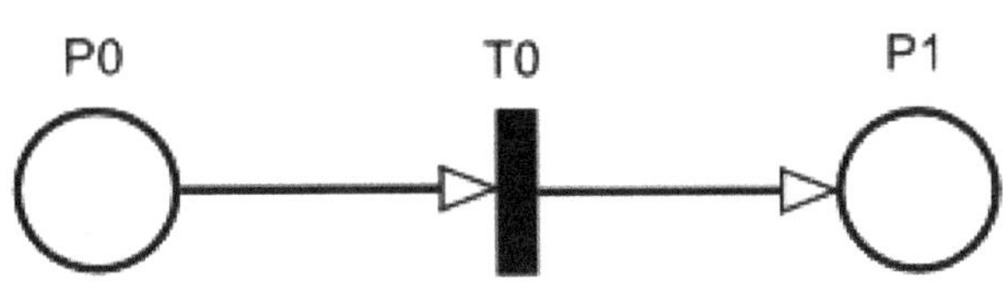

Figure 5. Example of Petri net.

Let M a place pi ∈ P, of a Petri net marked RM = (R, M0), this place is k-bounded (k ∈ IN) or simply limited if for every accessible marking M ∈ CA (R, M0), M (pi) ≤ k.

The limited k is the maximum number of marking that a place can accumulate. A Petri net labeled RM = (R, M0) is k-bounded if the number of marking at each RM site does not exceed k at any accessible RM marking (max (M (p)) = k, ∀ p ∈ P).

Safeness is a particularization of limited property. The concept of limited defines that a pi place is k-bounded if the number of marking that this place can accumulate is limited to the number k. A place that is 1-limited can simply be called insurance.

Liveness is defined according to the triggering possibilities of the transitions. A Petri net is considered live if, regardless of the marking that are reachable from M0, it is always possible to trigger any transition of the Petri net through a sequence of transitions L(M0). The absence of deadlock in systems is strongly linked to the concept of vivacity, since deadlock in a Petri net is the impossibility of triggering any transition of the Petri net. The fact that a system is deadlock free does not mean that it is live, however a live system implies a deadlock free system.

The concept of coverage is associated with the concept of reachability and live. An Mi marking is covered if there is a marking Mj ≠ Mi, such that Mj ≥ Mi.

6.1.2. Structural properties

The structural properties are those that depend only on the structure of the Petri net. These properties reflect independent marking characteristics. The properties analyzed in this work are structural limitation and consistency.

A Petri net R = (P, T, F, W, M0) is classified as structurally limited if it is limited to any initial marking.

The Petri net is considered to be consistent if, by triggering a sequence of enabled transitions from an M0 marking, it returns to M0, however all transitions of the Petri net are fired at least once.

Let RM = (R, M0) be a marked Petri net and a sequence s of transitions, RM is consistent if M0 [s > M0] and every transition Ti, firing at least once in s.

6.2. Stochastic Petri net

Petri Net (SNP) [11] is one of the Petri net extensions (PN) [15] used for performance and dependability modeling. A stochastic Petri net adds time to Petri net formalism, with the difference that the times associated with the timed transitions are exponentially distributed, while the time associated with the immediate transitions is zero. The timed transitions model activities through the associated times, so that the timing transition period corresponds to the activity execution period, and the timed transition trigger corresponds to the end of the activity. Different levels of priority can be assigned to transitions. The trigger priority of the immediate transitions is higher than the timed transitions. Priorities can solve situations of confusion [12]. The firing probabilities associated with immediate transitions can resolve conflict situations [4, 5].

Timed transitions can be characterized by different memory policies such as Resampling, Enabling memory and Age memory [5]. The timed transitions can also be characterized by different firing semantics named single server, multiple server and infinite server [5].

6.3. Phase approximation technique

SPN models consider only immediate transitions and timed transitions with exponentially distributed trigger times. These transitions model actions, activities, and events. A variety of activities can be modeled through the use of constructor throughput subnets and s-transitions. These constructs are used to represent expolinomial distributions, such as the Erlang, hypoexponential and hyperexponential distributions [9].

The phase approximation technique can be applied to model non-exponential actions, activities, and events through moment matching. The presented method calculates the first moment around the origin (average) and the second central moment (variance) and estimates the respective moments of the s-transition [9].

Performance and dependability data measured or obtained from a system (empirical distribution) with mean μ and standard deviation σ may have their approximate stochastic behavior through the phase approximation technique. The inverse of the variation coefficient of the data measured or obtained from a system Eq. (11) allows the selection of the expolinomial distribution that best adapts to the empirical distribution. This empirical distribution can be continuous or discrete. Among the continuous distributions, there are: Normal, Lognormal, Weibull, Gamma, Continuous Uniform, Pareto, Beta and Triangular and among the discrete distributions there are: Geometric, Poisson and Discrete Uniform [8].

$$\frac{1}{CV} = \frac{\mu_D}{\sigma_D} \tag{11}$$

The Petri net described in **Figure 6** represents a timed activity with generic probability distribution.

Depending on the inverse of the variation coefficient of the measured data (Eq. (11)), the respective activity has one of these distributions attributed: Erlang, Hypoexponential or Hyperexponential. When the inverse of the variation coefficient is an integer and different from one, the data must be characterized by the Erlang distribution. When the inverse of the variation coefficient is a number greater than one (but not an integer), the data are represented by the hypoexponential distribution. When the inverse of the variation coefficient is a number smaller than one, the data must be characterized by a hyperexponential distribution.

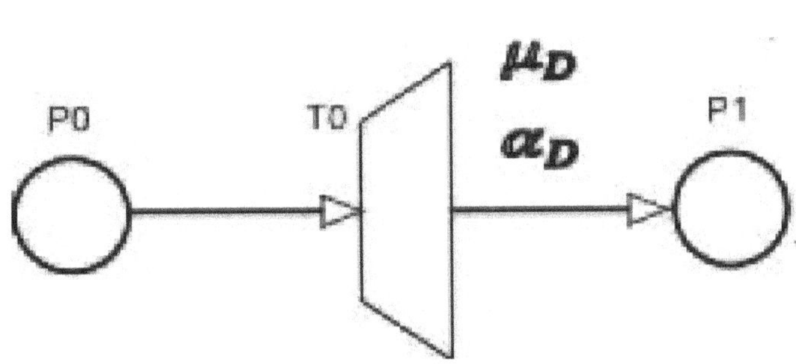

Figure 6. Empirical distribution.

7. Modeling strategy

The dependability metrics can be calculated using state space-based models (e.g., SPN) and combinatorial models (e.g., RBD). The RBDs have an advantage over the provision of results, as present faster calculations through their formulas than the simulations and the numerical analyzes of the SPNs. However, SPNs have a greater power of representation [3, 17].

State space-based models can describe dependencies that allow the representation of complex redundancy mechanisms. However, these models can generate a very large or even infinite number of states when they represent highly complex systems [3, 12, 17].

The combination of state space-based models and combinatorial models allows for the reduction of complexity in the representation of systems. RBD models can represent the components of the cloud computing [6]. These RBD models are used to estimate the availability and downtime of the cloud computing when there is little dependency relation between the components of this environment and the redundancy mechanisms adopted. If there is a need to represent a greater dependency between the components of the cloud computing and the redundancy mechanisms used, SPN models are used to represent the computational cloud systems [11].

Figure 7 shows a basic SPN model that allows a representation of the cloud computing. In this SPN model, the ON and OFF places represent the working or faulted computational cloud. The attributes of the transitions of this SPN model are presented in **Table 1**.

Figure 8 shows a basic RBD model that allows a representation of the cloud computing. The parameters of the RBD model are presented in the **Table 2**.

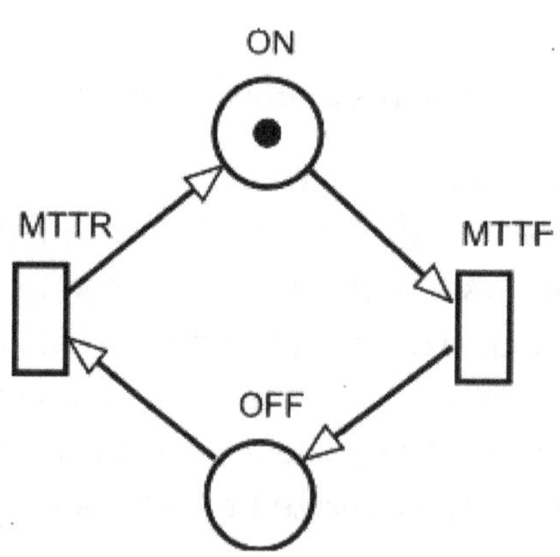

Figure 7. Basic SPN model.

Transition	Type	Time	Weight	Concurrence
MTTF	exp	XMTTF	–	SS
MTTR	exp	XMTTR	–	SS

Table 1. Attributes of the SPN model transitions.

Figure 8. Basic RBD model.

Parameters	Description
MTTFBlock	Mean Time to Failure
MTTRBlock	Mean Time to Repair

Table 2. Parameters of the RBD model.

7.1. Cloud computing model

Cloud computing consists of the Cloud Controller (Controller), Node Controller (Node) and Network equipment (Network). **Figure 9** shows the RBD model of the cloud computing. The parameters of the RBD model of cloud computing are presented in **Table 3**. **Figure 10** shows the SPN model of the cloud computing. The attributes of the SPN Model Transitions of cloud computing are presented in **Table 4**. In RBD and SPN models, the cloud controller and node controller are configured on different physical machines. The node controller enables the instantiation of virtual machines. The physical machines where the components of cloud computing are configured are connected through a switch and a router. All components of cloud computing must be operational for the cloud computing to be operational. These components can be described as Controller, Node, and Network. In this way, the operating mode of cloud computing is OM = (Controller ∧ Node ∧ Network).

Cloud computing consists of the Cloud Controller (Controller), Node Controller (Node) and Network equipment (Network). The Cloud Controller (Controller) has a hot standby redundancy, but the other components (Node and Network) can also be assigned this redundancy. The main cloud controller (ControllerMain) and the redundant cloud controller (ControllerStandby) in hot standby are operational [3, 4]. The operating mode of cloud computing with redundant cloud

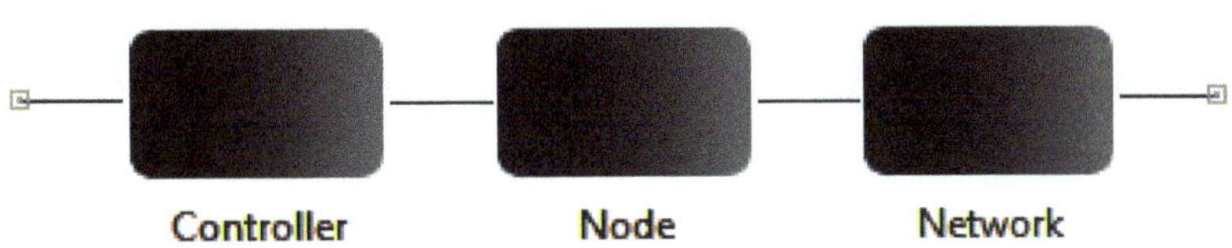

Figure 9. RBD model of the cloud computing.

Parameters	Description
MTTFController, MTTFNode, MTTFNetwork	Mean Time to Failure of the controller, node and network
MTTRController, MTTRNode, MTTRNetwork	Mean Time to Repair of the controller, node and network

Table 3. Parameters of the RBD model of the cloud computing.

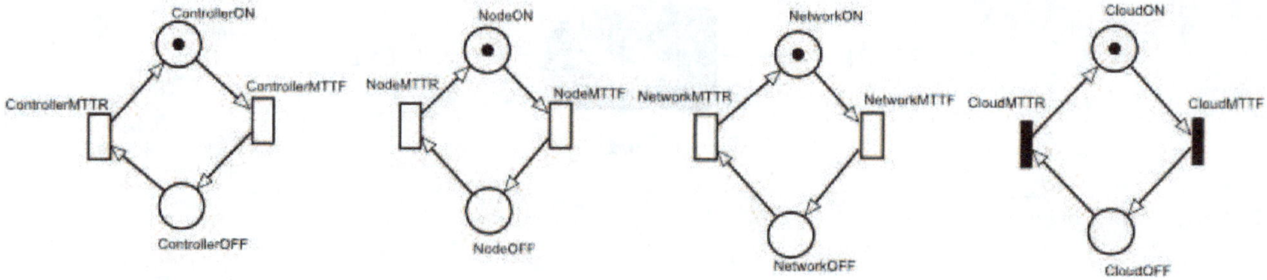

Figure 10. SPN model of the cloud computing.

Transition	Type	Time	Weight	Concurrence	Enable function
ControllerMTTF, NodeMTTF, NetworkMTTF	exp	XMTTF	–	SS	–
ControllerMTTR, NodeMTTR, NetworMTTR	exp	XMTTR	–	SS	–
CloudMTTF	imme	–	1	–	((#ControllerON = 0)OR(#NodeON = 0)OR (#NetworkON = 0))
CloudMTTR	imme	–	1	–	NOT((#ControllerON = 0)OR(#NodeON = 0) OR(#NetworkON = 0))

Table 4. Attributes of the SPN model transitions of the cloud computing.

controller in hot standby is OM = ((ControllerMain ∨ ControllerStandby) Λ Node Λ Network)). **Figure 11** shows the RBD model adopted to estimate the availability of cloud computing with redundant cloud controller in hot standby.

Cloud computing consists of the Cloud Controller (Controller), Node Controller (Node) and Network equipment (Network). The Cloud Controller (Controller) has a cold standby redundancy, but the other components (Node and Network) can also be assigned this redundancy. The main cloud controller (ControllerMain) is operational and the redundant cloud controller (ControllerStandby) is non-active. The redundant cloud controller is not operational waiting to

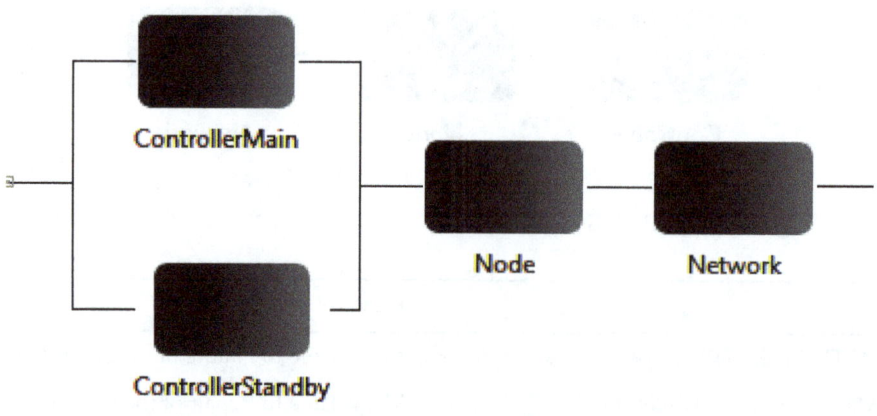

Figure 11. RBD model of the cloud computing with redundant cloud controller in hot standby.

be activated when the main cloud controller fails. Thus, when the main cloud controller, the activation of the redundant cloud controller occurs in a certain period of time. This period is named Mean Time to Active (MTA) [3, 4]. The operating mode of cloud computing with redundant cloud controller in cold standby is OM = ((ControllerMain ∨ ControllerStandby) Λ Node Λ Network)). **Figure 12** shows the SPN model adopted to estimate the availability of cloud computing with redundant cloud controller in cold standby.

Cloud computing consists of the Cloud Controller (Controller), Node Controller (Node) and Network equipment (Network). The Cloud Controller (Controller) has a warm standby redundancy, but the other components (Node and Network) can also be assigned this redundancy. The main cloud controller (ControllerMain) is based on a non-active redundant cloud controller (ControllerStandby) that waits to be activated when the main cloud controller fails. The difference with respect to cold standby redundancy is that the main cloud controller and the redundant cloud controller have an λ failure rate when they are in operation, but the redundant cloud controller has a failure rate φ when it is de-energized, considering that $0 \leq \lambda \leq \varphi$ [3, 4]. The redundant cloud controller (ControllerStandby) starts in idle mode. When the main cloud controller (ControllerMain) fails, the timed SpareActive transition triggers. This fire represents the start of the redundant cloud controller operation. The time associated with the SpareActive timed transition represents the Mean Time to Active (MTA). The SpareNActive immediate transition represents the return of the main module to the operational mode. The operating mode of cloud computing with redundant cloud controller in warm standby is OM = ((ControllerMain ∨ ControllerStandby) Λ Node Λ Network)). **Figure 13** shows the SPN model adopted to estimate the availability of cloud computing with redundant cloud controller in warm standby.

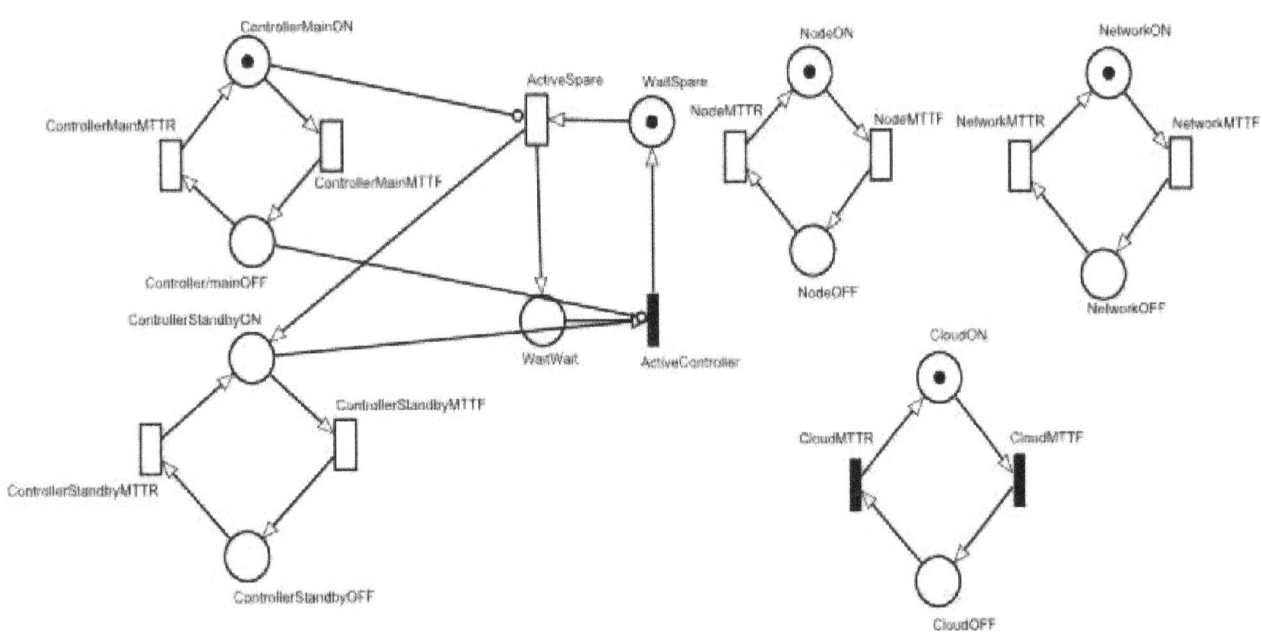

Figure 12. SPN model of the cloud computing with redundant cloud controller in cold standby.

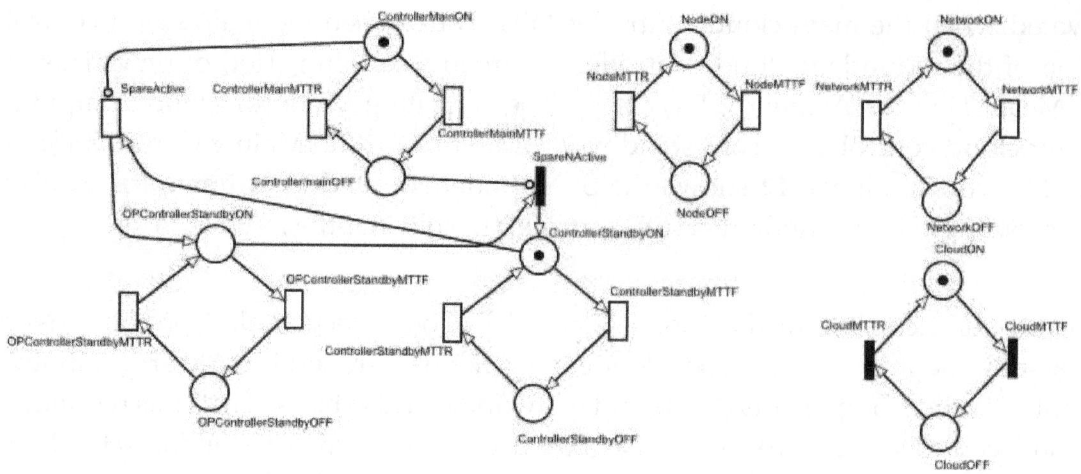

Figure 13. SPN model of the cloud computing with redundant cloud controller in warm standby.

8. Conclusions

This chapter presents concepts on dependability, redundancy mechanisms, stochastic Petri nets and reliability block diagram. In addition, this chapter also shows how the mathematical formalisms stochastic Petri nets and reliability block diagrams can be adopted for modeling cloud infrastructures with cold standby, warm standby and hot standby redundancy mechanisms. Reliability block diagrams is adopted to model cloud infrastructures with the redundancy mechanism cold standby and stochastic Petri nets is used to model cloud infrastructures with the redundancy mechanisms warm standby and hot standby.

Author details

Erica Teixeira Gomes de Sousa* and Fernando Antonio Aires Lins

*Address all correspondence to: erica.sousa@ufrpe.br

Department of Statistics and Informatics, Federal Rural University of Pernambuco, Brazil

References

[1] Bauer E, Adams R. Reliability and Availability of Cloud Computing. Wiley Online Library; 2012

[2] Laprie JCC, Avizienis A, Kopetz H. Dependability: Basic Concepts and Terminology. Secaucus, NJ, USA: Springer-Verlag New York, Inc; 1992

[3] Kuo W, Zuo MJ. Optimal Reliability Modeling: Principles and Applications. Wiley; 2002

[4] Rupe JW. Reliability of computer systems and networks fault tolerance, analysis, and design. IIE Transactions. 2003;**35**(6):586-587

[5] Maciel P, Trivedi K, Matias R, Kim D. Performance and Dependability in Service Computing: Concepts, Techniques and Research Directions. IGI Global; 2011

[6] Xie M, Dai YS, Poh KL. Computing System Reliability: Models and Analysis. US: Springer; 2004

[7] Ebeling CE. An Introduction to Reliability and Maintainability Engineering. Waveland Pr Inc; 2009

[8] Schmidt K. High Availability and Disaster Recovery: Concepts, Design, Implementation. Vol. 22. Berlin Heidelberg: Springer-Verlag; 2006

[9] Sahner RA, Trivedi K, Puliafito A. Performance and Reliability Analysis of Computer Systems: An Example-Based Approach Using the SHARPE Software Package. New York, US: Springer; 1996

[10] Trivedi KS. Probability & Statistics with Reliability, Queuing and Computer Science Applications. 2nd ed. Wiley; 2001

[11] German R. Performance Analysis of Communication Systems with Non-Markovian Stochastic Petri Nets. New York, NY, USA: John Wiley & Sons, Inc; 2000

[12] Marsan MA, Balbo G, Conte G, Donatelli S, Franceschinis G. Modelling with Generalized Stochastic Petri Nets, ACM SIGMETRICS Performance Evaluation Review. Vol. 26. New York, NY, USA; 1998

[13] Trivedi KS, Hunter S, Garg S, Fricks R. Reliability analysis techniques explored through a communication network example. Citeseer, International Workshop on Computer-Aided Design, Test, and Evaluation for Dependability; 1996

[14] Smith DJ. Reliability, Maintainability and Risk: Practical Methods for Engineers. Butterworth-Heinemann; 2011

[15] Murata T. Petri nets: Properties, analysis and applications. IEEE, Proceedings of the IEEE. 1989;**77**(4):541-580

[16] Maciel PRM, Lins RD, Cunha PRF. Introduction of the Petri Net and Applied. Campinas, SP: X Escola de Computação; 1996

[17] Balbo G. Introduction to Stochastic Petri Nets. Lectures on Formal Methods and Performance Analysis: First EEF/Euro Summer School on Trends in Computer Science. Berg en Dal, The Netherlands, July 3–7, 2000: Revised Lectures: Springer; 2000

5

Software Fault Injection: A Practical Perspective

Lena Feinbube, Lukas Pirl and Andreas Polze

Abstract

Software fault injection (SFI) is an acknowledged method for assessing the dependability of software systems. After reviewing the state-of-the-art of SFI, we address the challenge of integrating it deeper into software development practice. We present a well-defined development methodology incorporating SFI—fault injection driven development (FIDD)—which begins by systematically constructing a dependability and failure cause model, from which relevant injection techniques, points, and campaigns are derived. We discuss possibilities and challenges for the end-to-end automation of such campaigns. The suggested approach can substantially improve the accessibility of dependability assessment in everyday software engineering practice.

Keywords: fault injection, dependability, fault tolerance, testing, test-driven development

1. Introduction

On 22 October 2012, a major service degradation at Amazon Web Services (AWS)[1] affected several popular online services for several hours. It was caused by a latent memory leak bug, activated under stress due to a failed domain name system (DNS) update after a hardware maintenance event. The leaky software agent repeatedly tried in vain to contact the replaced server. In this process, memory was leaked until customer requests could no longer be handled.

AWS is a system so complex that it challenges exhaustive formal verification. The issue could, however, have been anticipated by structured experimental dependability assessment, e.g., using fault injection. Indeed, Netflix customers remained unaffected.[2] Resiliency testing had prepared the company for such events, and failover of Netflix servers worked quickly.

[1]http://aws.amazon.com/de/message/680342/

[2]http://techblog.netflix.com/2012/10/post-mortem-of-october-222012-aws.html

This incident highlights a central issue with the current state of dependability research. There is a vast gap between the impressive theoretical achievements made in past decades, including various formal methods, and real-world software engineering practice.

Similar incidents suggest that efforts to increase systems' dependability should be shifted toward the software layers· As complexity grows, these concerns are becoming ever more challenging.

2. Fundamentals

Software dependability research is as old as the science of computing itself. In the early nineteenth century, Charles Babbage designed his famous difference engines with the main intent of reducing the error rate in complex mathematical computations [1].

As we have started to rely on software in many safety-critical aspects of our daily lives, software dependability is more relevant than ever. At the same time, the problem is getting harder. If the number of flaws (bugs) per source code lines is roughly constant, software projects increasing in code size would become more and more prone to failures. This is aggravated by the complexity caused by the interaction of different components.

Therefore, software dependability needs to take a holistic, system-wide approach, and dependability means need to scale to increasingly complex software projects. This includes means for fault forecasting and dependability assessment, which are the focus of this chapter.

2.1. Terminology

Software dependability has been characterized by a broad range of terminology [2]. We employ the wording by Avižienis, Kanoun, Kopetz, Landwehr, Laprie, and Randell, described in multiple documents dated between 1985 and 2004. The most comprehensive version [3] is used as main reference. In this terminology, dependability is threatened first by *faults*:

> *A fault is the adjudged or hypothesized cause of an error.*

In software, the terms *bug*, *defect*, or *flaw* are often used as synonyms for *fault*. When activated, a fault can lead to an undesired system state, denoted as *error*:

> *An error is that part of the system state that may cause a subsequent failure.*

There is an n:m relationship between faults and their resulting error states. A fault, when activated under varying environmental and internal conditions, might lead to different error states. In turn, each error state might be caused by different—potentially multiple—faults. Finally, error states may lead to an externally visible *failure*:

> *A system failure is an event that occurs when the delivered service deviates from correct service.*

2.2. Dependability assessment

How can the dependability of a complex software system be assessed and compared with other versions or other systems?

The complexity of software is caused by three factors:

1. The internal state space of the program, given by all variables and their values.
2. The space of input and output values of the program, which may be infinite.
3. Interaction with the environment, which is influenced, for instance, by scheduling, resource states, and interfaces to other software components.

These complexities challenge the scalability of any dependability assessment approach. There are two ends to the spectrum of such methodologies, which we classify as *formal methods* and *empirical methods*. While the spectrum is continuous and the boundaries may be fuzzy, the main differences lie in their coverage, in the assets they use (static source code vs. runtime information), and in the scalability to large software systems.

The aim of formal methods is to exhaustively prove that a program (an implementation) obeys a specification. Various efforts toward complete formal software verification have been made despite theoretical limitations as well as scalability issues.

Empirical methods for software dependability take the approach of showing that a software system operates correctly at the example of one or many executions of the program. While such approaches never provide absolute guarantees, they are generally better scalable and applicable to larger, more complex, and rapidly evolving systems. Of course, as the often-cited Dijkstra famously noted,[3]

> *Program testing can be used to show the presence of bugs, but never to show their absence!*

2.3. Software fault injection

Software fault injection (*SFI*) denotes the artificial insertion—*injection*—of faults and error states into a running software system. It can be applied beyond the limits of formal verification methods because it does not assume a complete formal specification of the system under test. The experimental approach of SFI can be implemented efficiently and with little intrusiveness.

In the simplest case, SFI can confront an interface with randomly generated values. More sophisticated SFI operates based on detailed failure cause models and can rely on formal specifications to work more efficiently and to guarantee certain fault space coverage criteria. SFI tools can compare the dependability of different systems and answer research questions such as:

- How high is the coverage of fault tolerance mechanisms and how well do they work?
- How do faults influence quality metrics such as performance, precision, or availability?
- Are there single points of failure in the system?

Fault injection can be considered a complimentary technique to software testing. While software tests are designed to assert correct behavior of the system under a representative workload, fault injection asserts correct behavior under an additional *faultload*.

[3]https://www.cs.utexas.edu/users/EWD/transcriptions/EWD03xx/EWD303.html

3. Software fault injection—state-of-the-art

Fault injection is a versatile tool for dependability assessment. When an injected fault causes a system failure, this can indicate insufficient fault tolerance mechanisms. In general, the system can only be assumed robust and dependable when all its layers and components are fault tolerant and their fault tolerance has been verified empirically or formally.

Various fault injection implementation strategies with different characteristics exist. One classification thereof was given in Ref. [4] and is presented here in an adapted form.

Any fault injection tool relies on a **trigger mechanism** that causes the artificially generated fault or error to be inserted into normal program execution:

- *Time-based*: fault injection takes place at predetermined time intervals.
- *Location-based*: faulty values are written into predefined memory locations.
- *Execution-driven*: fault injection occurs dynamically, depending on the control flow.

While time-based fault injection can often easily be implemented non-intrusively, this is not the case for location-based fault injection, which is suitable for memory corruption, but impedes controlling the fault load dynamically. Execution-based fault injection allows for complex and realistic fault models, but is also not applicable to black box applications.

SFI can take place at different **injection times**:

- *Before runtime*: the program is modified upfront, for instance, by using source code mutation to add faults (software bugs) to the code.
- *During runtime*: faults are injected during program execution.
- *At the loading time of external components*: here, injection triggers may be the dynamic binding of external libraries or the adding of other dependencies during runtime.

Further, we distinguish between *synchronous* triggers, such as exception handling mechanisms, and *asynchronous* triggers, such as hardware interrupts.

Another question is which representations of a program, called **injection artifacts**, are to modify for fault injection purposes:

- Machine language or binary files
- Intermediate representations of the source code
- Source code

In general, as with other testing and profiling approaches, the behavior of a software system inevitably changes under faultload. To minimize this change, approaches should be as little intrusive as possible. The questions which fault injection approach to use, when to, and where to apply it using which faultload determine the effectiveness and efficiency of SFI.

3.1. Software fault injection approaches

A comprehensive survey of SFI techniques has been presented by Natella et al. [5]. This section presents selected SFI tools at different layers of abstraction within the software stack.

3.1.1. *Operating system-level fault injection*

The tool *Ballista* [6] has been used to compare different OS' dependability by exercising system calls with both valid and invalid parameters. It bridges the gap between software testing and SFI by using test input selection methods, based on coverage criteria, to choose adequate fault injection targets.

Crashme[4] is a simple tool that tests the robustness of operating systems by attempting to execute random byte sequences as procedures. Despite its simplistic nature, crashme uncovered severe flaws in different operating systems and hypervisors.

Trinity[5] was a widely used "fuzzer"—i.e., a tool generating randomized inputs—for Linux system call interfaces, incorporating knowledge about file pointers and other OS-specific resource descriptors. The tool has been used to find many bugs in the Linux kernel.

3.1.2. *Fault injection at interfaces*

Targeting interfaces between software libraries, the SFI tool suite *execution driven fault injection (EDFI)* [4], enables the definition of fine-grained fault loads and extensible dynamic fault triggers. It relies on a combination of dynamic and static source code instrumentation.

Library fault injector (LFI) [7] injects faults at the interface between an application and its dynamically linked libraries. It applies binary analysis to obtain a fault model, describing which values and error codes can be returned from a C-style library and to analyze whether they are handled.

Further language-specific libraries for testing the robustness against interface faults exist, many of which are included in language runtimes, such as *Libfiu*[6] and Microsoft *TestApi.*[7]

3.1.3. *Fault injection in distributed systems*

In distributed systems, fault injection has a long history, mainly considering hardware and message passing in their failure cause models.

Orchestra [8] defines an architecture for inserting a protocol fault injection layer in the network stack, which applies filters to messages sent and received, to inject faults into them.

Chaosmonkey,[8] which has evolved into the *SimianArmy* resiliency testing tool suite, targets applications built upon AWS by making single-node instances unavailable. The approach of

[4]http://people.delphiforums.com/gjc/crashme.html
[5]https://codemonkey.org.uk/projects/trinity/
[6]https://blitiri.com.ar/p/libfiu/
[7]https://testapi.codeplex.com/
[8]https://github.com/Netflix/SimianArmy/wiki/Chaos-Monkey

injecting faults into deployed systems was first successfully employed by Netflix and has been adapted by other companies.

With the advent of cloud computing, there has been a paradigm shift from trying to avoid failures at all costs to embracing faults as opportunities for making the system more resilient. The amount of effort put into the fault tolerance and resilience of cloud applications is often determined by service level agreements (SLAs), and the trade-offs between development effort, costs for redundancy, availability, and consistency are usually application-specific and based on management decisions.

3.2. Software failure cause models

Software failures are complex in their nature, origins, and manifestations. For SFI, a detailed understanding of the causality chains leading to software failure is needed. Software failure causes have been studied in various domains of software engineering research, providing a broad and heterogeneous spectrum of published models.

In a systematic literature study [2], we found that there is a lack of dynamic fault activation and error models. Such models are especially relevant for SFI. As shown in **Figure 1**, most of the papers—123 out of 156 publications—discuss a static, code-based fault model.

As software is becoming more and more crucial to dependability, so are software-focused failure cause models. Recent efforts to standardize knowledge about software failure causes include orthogonal defect classification (ODC) [9] and the common weakness enumeration (CWE) database.[9]

Additional layers of abstraction, as introduced, for example, by cloud management stacks, call for novel and more targeted failure cause models.

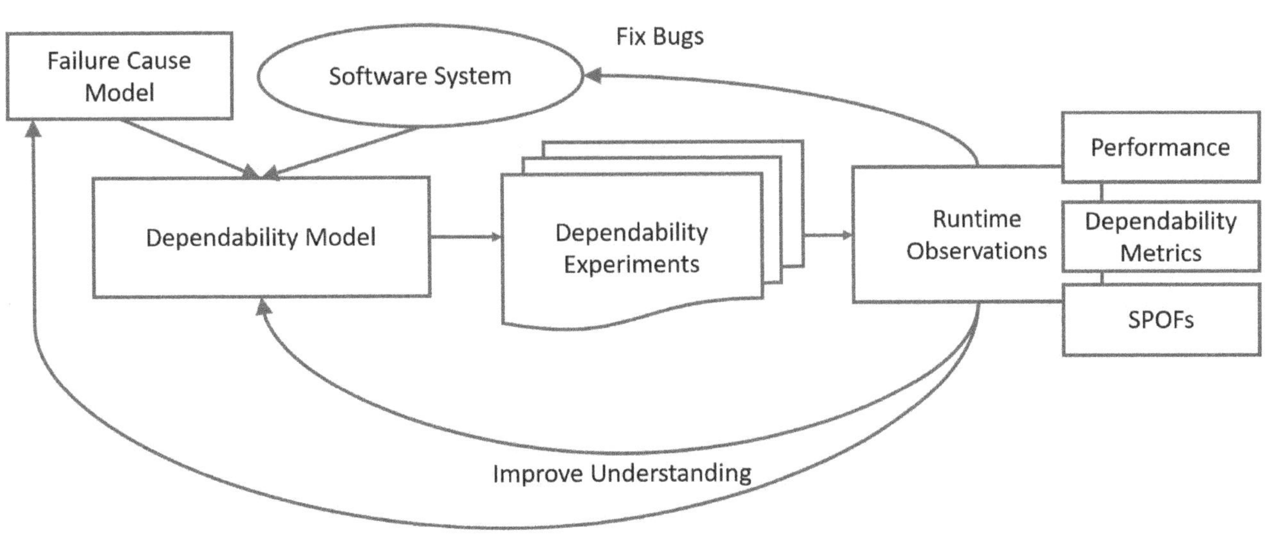

Figure 1. Literature study of failure cause models—most research targets static aspects.

[9]https://cwe.mitre.org

4. Problem statement

SFI is a versatile tool for dependability assessment and for testing the fault tolerance of increasingly complex systems. However, it is not yet widely used in real-world software systems outside the safety-critical and embedded domains. There is a gap between comparatively formal approaches, such as model-driven engineering, and what we choose to call *ad hoc software engineering*.

Formal approaches suffer from several limitations. They are successful in rather constrained domains, where full formal specification and fault modeling are possible. They are generally not applicable in modern scenarios involving rapidly changing requirements, under specified execution environments and fast-evolving software dependencies.

On the other hand, ad hoc software engineering is limited to implementing a software system according to some partial and informal specification. The resulting product is tested by its developers—usually without making strong coverage guarantees—and released without little other dependability assessment.

We assume that ad hoc software engineering constitutes most state-of-the-art software systems and is currently the most pragmatic approach in many situations, where fully formalized approaches are out of scope for technical as well as organizational reasons.

The following sections suggest how to improve ad hoc software engineering explicitly considering dependability aspects and by making software fault injection a pivotal dependability assessment means throughout the development process.

5. Methodology

We introduce *Fault Injection Driven Development* (*FIDD*), depicted in **Figure 2**, which integrates SFI into the development process. The following sections elaborate on its details.

5.1. Creating a dependability and failure cause model

To make the faultload introduced by SFI representative, two aspects need to be understood: The *dependability model* should capture the intended behavior in the presence of faults. This includes redundancy configurations, as well as error detection and recovery mechanisms. The *failure cause model* contains faults, errors, and fault activation conditions [10] that are anticipated and somehow addressed in the system architecture.

Modeling can serve to understand a complex system's dependability aspects to qualitatively discuss dependability bottlenecks or to quantitatively evaluate dependability metrics. Numerous modeling languages with varying expressiveness have been proposed. Here, we distinguish between *structural* and *behavioral* modeling languages.

Structural languages reflect the system structure typically at a higher level of abstraction such as components. The interactions and error propagation between components are in the

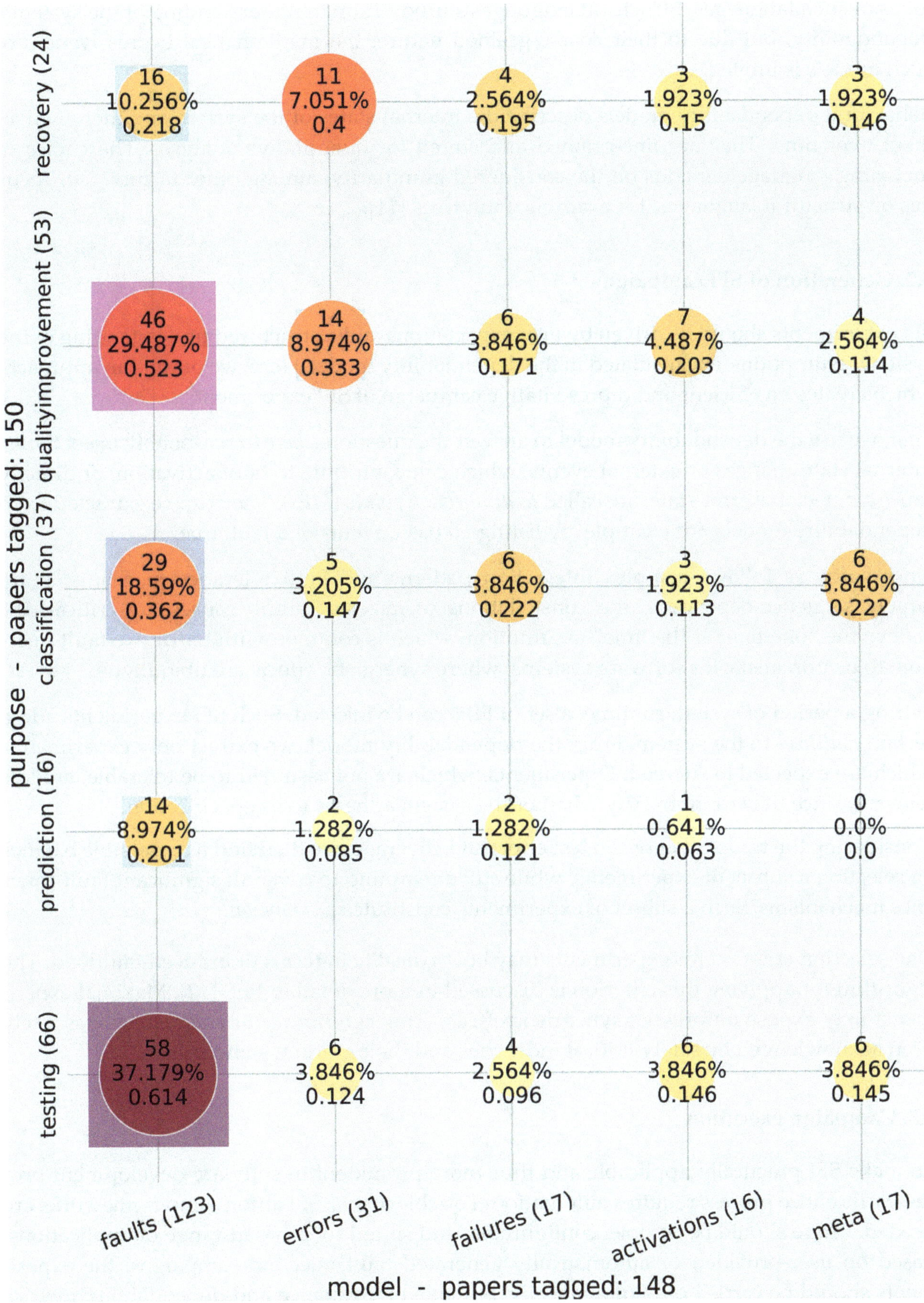

Figure 2. Overview of our fault injection–driven development methodology.

focus of such languages. Structural languages support human understanding of the system's dependability, but due to their coarse-grained nature, the mathematical expressiveness of such models is limited.

Behavioral dependability models describe the internal states of the system and their evolvement over time. They are fine-grained and suited for quantitative analyses. The choice of modeling language depends on the considered granularity, among other factors. Our focus lies on structural languages, for example, fault trees [11].

5.2. Generation of SFI campaigns

SFI experiments should be driven by user expectations and a structured understanding of the system assumptions, encapsulated in the dependability model. Here, we outline an approach, which creates an efficient and representative campaign of SFI experiments.

First, we use the dependability model to answer the question *where* to reasonably inject faults. Internal state changes or external events, which can contribute to fault activation, or directly cause a detectable error state, are called *fault injection points* (*FIPs*). They can be extracted from dependability models, for example, by listing all basic events of a fault tree.

Since software follows complex interaction patterns and fault tolerance mechanisms are largely sequence-dependent, the constellations of injected failure causes are critical. We believe the "one fault at the time" assumption, which is common with hardware fault injection, to be unrealistic for software systems, where synergistic effects are ubiquitous.

During a period of system runtime, a set of FIPs can be injected. Such SFI experiments add a certain *faultload* to the system. From the dependability model, we extract only experiments, which are expected to succeed. Experiments, which are not assumed to be tolerable, are less relevant, since they cannot verify whether the system adheres to its specification.

Considering the trade-off between test effort and coverage, we then find a reasonable balance by selecting a subset of experiments, while still attempting to cover all significant fault-tolerance mechanisms. Such a subset of experiments constitutes a *campaign*.

One selection criterion for experiments may be *maximality* in terms of induced faultload. The algorithm for applying this criterion is discussed in more detail in Ref. [12]. Maximal experiments may expose unforeseen synergistic effects. The significance of such effects has lately been acknowledged by safety-critical industries and their guiding standards [13].

5.3. Campaign execution

To make SFI practically applicable and thus more prominent in software development processes, its entire process requires automation. To achieve this, SFI automation frameworks are needed. These should be flexible, configurable, and suited for a broad range of applications. Based on user-provided or automatically generated fault injection campaigns, the experiments should be carried out automatically. Suitable performance and dependability metrics for analysis should be gathered.

In a previous work, we have presented two such SFI frameworks at different layers of abstraction:

Hovac [14] is an SFI framework for multithreaded applications, implemented in C++, which enables the customizable injection of various faults by using application programming interface (API) hooking, a nonintrusive technique requiring no source code access. The implemented faults and error states are based on the community-based CWE database.

We are currently also working on a tool suite, Faultmill, which fully automates the generation and orchestration of SFI campaigns in distributed systems. It relies on user-provided dependability models and provides a user-friendly interface for integrating custom fault injection scripts and executables.

5.4. Analysis and feedback

A remaining open question is which conclusions to draw from the results of campaigns, and how they can be leveraged to improve the overall process. If all experiments in the campaign succeed, this validates the initial dependability model. Furthermore, by gathering quality metrics, such as runtime or accuracy, the degradation under faultload can be quantified.

Known from software testing, *coverage measures* provide a quality metric for the assessment process itself. In a nutshell, they express how well a set of experiments (unit tests, SFI campaigns, or others) covers the vast space of possible behaviors. To provide feedback on the campaign, we suggest applying structural coverage criteria on the model.

6. Case study: SFI contest

The remainder of this chapter discusses a concise case study of applying SFI in software development teams: an SFI contest was carried out in a student seminar. Within the contest, we explored approaches and challenges of FIDD within a software development team. The contest took place during one semester within a course for 6 ECTS.

image of the Julia fractal[10] in parallel using OpenMP.[11]

were planning to implement in their fault injector. Subsequently, they were asked to specify the intended behavior of their own implementation under faultload.

The fault injection was implemented in a virtual machine (VM) running a (modified) Ubuntu Linux. VMs were configured using Vagrant[12] for portability. A fuzzing script was developed by us, which executed the students' fault-tolerant program within the fault-injecting VM environment, providing corner case and randomly generated input arguments.

[10]http://mathworld.wolfram.com/JuliaSet.html

[11]http://www.openmp.org/

[12]https://www.vagrantup.com/docs/getting-started/

6.1. Failure cause and dependability model

Each team first developed a *failure cause model* to be assessed with SFI. This process required some coordination between the teams to ensure that failure causes were on similar layers of abstraction. The resulting classes of failure cause are customized to a multithreaded C/C++ application running in a potentially unreliable execution environment:

- **Incorrect arguments**: unexpected parameters are passed to function calls or interfaces.
- **Memory errors**: the allocation of memory may fail or the memory may be corrupted.
- **Resource scarcity**: resources such as CPU time and file handles may be limited as well.
- **I/O errors**: I/O operations may be unreliable—they may fail, stall, or return errors.
- **Data type errors**: assumptions made on the data type or layout may be violated.
- **Erroneous system time**: the current time, queried by a system call, may be incorrect. This simulates time lags in distributed systems, as well as software or hardware flaws.

The intended system behavior under faultload was specified by both teams and took the failure cause model into consideration. While (due to time constraints) the specifications were not completely formalized but rather formulated in natural language, they offer some notable insights.

First, the specification contained *quantitative aspects*, such as "if dynamic memory allocation fails more than five times, we assume a permanent error, which is not tolerable." Moreover, many assumptions on the environment were made. This stresses the importance of environment-aware failure cause models. Finally, little focus was placed by students on performance and other quality metrics. This may indicate the necessity of more research and education regarding trade-offs among nonfunctional software properties.

6.2. Implementation of SFI

The developed SFI approaches were based on the mentioned assumptions:

System calls were intercepted using the *LD_PRELOAD*[13] environment variable. This variable was used to enforce the loading of SFI libraries, which offer identical interfaces to some function calls, such as *malloc*. Failing I/O operations were simulated analogously. To tolerate such injections, some critical functions, including the *new*-operator for C++, were re-implemented by the students.

The scarcity of resources, for example caused by other processes on the system, was injected by using the *cgroups*[14] facility. The application was limited in its CPU shares and memory. A second *cgroup* containing resource-intensive processes was created to put the application under stress. One group addressed this issue by snapshotting state to hard disk periodically.

[13]http://man7.org/linux/man-pages/man8/ld.so.8.html

[14]https://www.kernel.org/doc/Documentation/cgroup-v1/cgroups.txt

In some scientific and distributed computations, an incorrect system time can lead to errors. Although this was not the case for the example application, the system time was manipulated using *libfaketime*.[15]

Bit-level memory errors were implemented by halting the program using *ptrace* and manipulating random memory areas. This injection of low-level faults turned out to be hard to predict and tolerate—it was addressed with redundant, checkpointed computations.

6.3. Lessons learned

In the contest, the most fault-tolerant application started five redundant processes, each checkpointing its state in short intervals. Under faultload, this application became so slow that the test scripts which were running it timed out.

There is an inherent trade-off between performance and fault tolerance, which induces runtime overhead for error detection, containment, and removal. This trade-off needs further analysis and specification.

In practice, the continuous integration of the SFI environment (a VM, submitted by the students) with the program template required a surprising amount of manual work due to software dependencies and hidden environmental assumptions. As full automation of SFI campaigns remains an engineering challenge, standardization of SFI interfaces is desirable.

As already mentioned, developing a dependability model and a failure cause model required a substantial amount of communication among the teams. While several efforts have been made, such as ODC and the CWE database, there is yet no established failure cause model, which is commonly known and used among software developers. Such a model would have been a powerful communication tool in the contest.

7. Discussion and conclusion

Software dependability is harder to ensure in complex systems with many layers of abstraction, interacting components, concurrency, and increased distribution.

SFI is a promising approach for evaluating the dependability of software systems, which scales even for large and complex systems, as the success of ChaosMonkey exemplifies. It is especially suited for ad hoc software engineering, where a formal understanding of the system is missing. Yet, SFI is not yet established in general software development practice.

We have introduced *Fault Injection Driven Development* (*FIDD*), a structured approach for applying SFI. We discuss how even starting without any specification or documented fault model, it is possible to make the dependability aspects explicit and create a fully automated environment, in which repeated fault injection campaigns assure dependability and yield relevant insights.

[15]https://github.com/wolfcw/libfaketime

The case study of an SFI student contest underlines the necessity of FIDD-like methodologies and provides insights into organizational and engineering challenges.

Many opportunities for further research remain: as discussed, most current failure cause models are not suited for SFI. They do not take dynamic aspects of software failures into account. However, dynamic fault activation and error models provide valuable input for SFI and are, therefore, topic of our ongoing research.

Practical experience shows that SFI frameworks are tedious to implement if they are to support full automation of campaign execution and extensibility. To support the reuse of such frameworks, some form of standardized interfaces for SFI would be immensely useful.

Acknowledgements

The authors would like to thank all participants of the student contest at HPI, during the winter semester 2016/17.

Author details

Lena Feinbube*, Lukas Pirl and Andreas Polze

*Address all correspondence to: lena.feinbube@hpi.de

Hasso-Plattner-Institute, University Potsdam, Germany

References

[1] Babbage C. Passages from the Life of a Philosopher. Longman, Green, Longman, Roberts, & Green; London; 1864

[2] Feinbube L, Tröger P, Polze A. The Landscape of Software Failure Cause Models. arXiv preprint arXiv:1603.04335. 2016

[3] Avižienis A, Laprie J-C, Randell B, Landwehr C. Basic concepts and taxonomy of dependable and secure computing. IEEE Transactions on Dependable and Secure Computing. 2004;**1**(1):11-33

[4] Giuffrida C, Kuijsten A, Tanenbaum AS. EDFI: A dependable fault injection tool for dependability benchmarking experiments. In: 2013 IEEE 19th Pacific Rim International Symposium on Dependable Computing (PRDC); 2013

[5] Natella R, Cotroneo D, Madeira HS. Assessing dependability with software fault injection: A survey. ACM Computing Surveys (CSUR). 2016;**48**(3):44

[6] Koopman P, Sung J, Dingman C, Siewiorek D, Marz T. Comparing operating systems using robustness benchmarks. In: Proceedings of the Sixteenth Symposium on Reliable Distributed Systems; 1997

[7] Marinescu PD, Candea G. LFI: A practical and general library-level fault injector. In: 2009 IEEE/IFIP International Conference on Dependable Systems & Networks; 2009

[8] Dawson S, Jahanian F, Mitton T. Orchestra: A Fault Injection Environment for Distributed Systems. In: Technical Report; 1996

[9] Chillarege R, Bhandari IS, Chaar JK, Halliday MJ, Moebus DS, Ray BK, Wong M-Y. Orthogonal defect classification-a concept for in-process measurements. IEEE Transactions on Software Engineering; 1992

[10] Tröger P, Feinbube L, Werner M. What activates a bug? A refinement of the Laprie terminology model. 2015 IEEE 26th International Symposium on Software Reliability Engineering (ISSRE); 2015

[11] Vesely WE, Goldberg FF, Roberts NH, Haasl DF. Fault Tree Handbook. DTIC Document; 1981

[12] Feinbube L, Pirl L, Tröger P, Polze A. Software fault injection campaign generation for cloud infrastructures. In: 2017 IEEE International Conference on Software Quality, Reliability and Security Companion; 2017

[13] International Organization for Standardization. Road vehicles—Functional safety. 2011; (ISO 26262:2011(E))

[14] Herscheid L, Richter D, Polze A. Hovac: A configurable fault injection framework for benchmarking the dependability of C/C++ applications. In: Proceedings of the 2015 International Conference on Software Quality, Reliability, and Security; 2015

6

Advances in Engineering Software for Multicore Systems

Ali Jannesari

Abstract

The vast amounts of data to be processed by today's applications demand higher computational power. To meet application requirements and achieve reasonable application performance, it becomes increasingly profitable, or even necessary, to exploit any available hardware parallelism. For both new and legacy applications, successful parallelization is often subject to high cost and price. This chapter proposes a set of methods that employ an optimistic semi-automatic approach, which enables programmers to exploit parallelism on modern hardware architectures. It provides a set of methods, including an LLVM-based tool, to help programmers identify the most promising parallelization targets and understand the key types of parallelism. The approach reduces the manual effort needed for parallelization. A contribution of this work is an efficient profiling method to determine the control and data dependences for performing parallelism discovery or other types of code analysis. Another contribution is a method for detecting code sections where parallel design patterns might be applicable and suggesting relevant code transformations. Our approach efficiently reports detailed runtime data dependences. It accurately identifies opportunities for parallelism and the appropriate type of parallelism to use as task-based or loop-based.

Keywords: parallelism, multicore/manycore systems, software engineering, code analysis, profiling

1. Introduction

Stagnating single core processor performance caused a new hardware trend in the past years that resulted in the replication of cores and the popularity and ubiquity of multicore and manycore architectures. Many applications and software systems that face growing demand for computational power can leverage this hardware trend for their needs and achieve reasonable performance via software parallelization. The only way for application developers to speed up an individual application is to match the new hardware cores with thread-level parallelism in the form of task-based or loop-based parallelism. However, successful parallelization is often error prone, difficult, and time-consuming, especially if it is done manually. Further, applying auto-parallelization is generally limited to loops with specific criteria, and it is based on the polyhedral model [1, 2] for compiler optimization. Additionally, auto-parallelization often fails to identify and exploit available parallelism for many applications, since it does not leverage runtime information such as pointers and array indices.

To keep application developers motivated and encourage them to achieve a performance improvement, automated tools and methods are necessary that support them during a semi-automatic parallelization process to reduce the manual efforts and facilitate the parallelization workflow. Hence, effective programming toolchain and methodologies for using an optimistic code-based approach to parallelize software with minimum programming effort and user intervention are in great demand.

There are three major problems in the software parallelization process that often cause the parallelization process to suffer from high complexity and low productivity. The first problem is gaining a thorough and complete understanding of the software code to identify detailed control and data dependences. In order to guarantee the program correctness, the parallelized program must have proper synchronization operations to preserve data dependences and the right order of data accesses to produce the same results as the sequential code does.

The second problem is extracting coarse-grained parallelism. Because of the available hardware parallelism in multicore/manycore processors, they are powerful in executing multiple code sections simultaneously. But the software programming toolchain is not mature to help programmers partition and map their code to the new available cores. Coarse-grained parallelism such as task-based parallelism is expected to be a promising solution for using the available hardware parallelism of the new cores and finding parallelism between arbitrary code sections.

The third problem is generating parallel code which can express this coarse-grained parallelism effectively for a diverse number of target platforms. After generating the parallel code, validation and verification will be applied, and for further performance improvements on specific targets, optimization techniques and auto-tuning methods are necessary.

This work summarizes the results of methods and approaches, which set out to improve the abovementioned problems. The main goal of the work was to make semiautomatic parallelization more feasible and attractive for a broader audience of application developers by providing tools and methods that use an optimistic code-based approach and support key activities of the manual parallelization process in a simpler, more effective and intuitive way than existing tools.

The remainder of the chapter is structured as follows: in the next section, we highlight the main contributions and the essential results of this work. In Section 3, we explain our approach to dependence profiling and decomposition. In Section 4, we briefly present our methods for task extraction and parallel pattern identification. Sections 5 and 6 deal with code transformation and correctness analysis, respectively. In Sections 7 and 8, several applications of our framework and its limitations are discussed. Section 9 reviews related work. Section 10 concludes the chapter and discusses possible extensions.

2. Contribution summary

The most important goal of this work is to provide a set of methods as an end-to-end solution to support programmers during the semiautomatic parallelization process, from the initial code analysis to code generation and optimization. The methods follow an optimistic code-based approach to effectively analyze different code sections based on the actual runtime dependence analysis. In this way, parallelization opportunities of applications can be identified at an early stage of the code analysis, which maximizes flexibility and also facilitates the parallelization process. The approach is implemented as a tool called Discovery of Potential Parallelism (DiscoPoP) [3] and is based on the LLVM compiler infrastructure.

The main accomplishments, which are illustrated in **Figure 1**, can be summarized as follows:

- **Dependence profiling.** We instrument and execute the program to obtain its control and data dependences with practical overhead [4]. Our data-dependence profiler serves as foundation for different program analyses based on data dependences.
- **Decomposition.** The concept of computational units (CUs) is used to extract the basic blocks for building parallel programs [5]. A CU follows the read-compute-write pattern, which means that a program state is first read from memory, a new state is computed, and finally the new state is written back to memory. We generate the CU graph of a program based on its CUs and the dependences that exist among them.
- **Task extraction and parallel pattern identification.** We search for potential parallelism in the program by merging CUs/partitioning the CU graph [3]. The output is a prioritized list of parallelization opportunities [6]. In a next step, we identify suitable parallel design patterns to support the parallel algorithm structure [7, 8].
- **Code transformation.** In simple cases, the program is automatically transformed into its parallel version based on available parallelism and the identified parallel design patterns [9]. In other cases, suggestions for parallelization are presented.
- **Correctness analysis.** Additionally, an automated method to generate unit tests targeting concurrency bugs such as data races has been developed to validate the resulting code [10, 11].
- Numerous applications and case studies, in which we confirm the functionality of our approach and show its capability as a parallelism discovery tool. In many cases, we can

reproduce manual parallelization strategies. We further demonstrate how DiscoPoP can serve as a basis for different kinds of program analyses such as the exposure of communication patterns [12] or the optimization of transactional memory [13, 14].

In the remainder of the chapter, we describe the above contributions in more detail, followed by a quick look at ongoing developments and a review of related work.

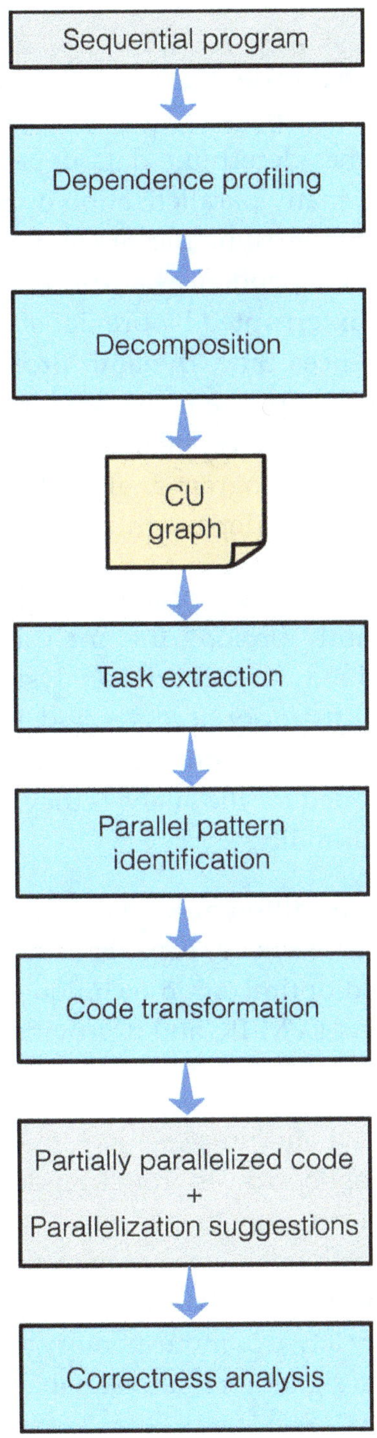

Figure 1. Main elements of the DiscoPoP parallelization approach.

3. Dependence profiling and decomposition

A key result of this work is a profiling method that reports data dependences of the executed program efficiently in terms of time and space. The reported data dependences are used to build the computational units that serve to analyze the program.

3.1. Dependence profiling

In order to parallelize the sequential code, we need to identify the control and data dependences of the program. Data dependences can be obtained in two major ways: static and dynamic analyses. Static approaches determine data dependences at compile time without executing the program. However, many parallelization opportunities are ignored due to the lack of runtime information. In contrast, dynamic dependence profiling instruments the intermediate or binary code and tracks dependences at runtime. It treats the execution of a user program as an instruction stream interrupted by previously inserted calls to instrumentation functions that help detect dependences. Since dynamic profiling tracks only the branches that are actually executed, it is inherently input sensitive, and it identifies control and data dependences for the actual program execution. Despite this, the results are still useful, which is why such profiling forms the basis of many program analysis tools. Moreover, by changing inputs and computing the union of all collected dependences, the input sensitivity can be mitigated.

However, a limitation of data-dependence profiling is high runtime and memory overhead. The time overhead may significantly prolong the analysis, sometimes requiring an entire night [15]. The memory overhead may prevent the analysis completely [16]. This is because dependence profiling requires all memory accesses and locations to be instrumented and recorded. To lower the overhead, current profiling approaches limit their scope to the subset of the dependence information needed for the analysis they have been created for. In this way, they reduce the generality and reusability.

To provide a general foundation for different kinds of analyses, we present a generic data-dependence profiler with practical overhead, capable of supporting a broad range of dependence-based program analysis and optimization techniques for both sequential and parallel programs. The profiler is based on LLVM-IR, and it provides detailed information, including source-code location, variable name, and thread ID.

The proposed profiler is parallelized and utilizes a lock-free design [17] to achieve efficiency. It leverages signatures [18], a concept borrowed from transactional memory to reduce memory consumption. A signature is a data structure that encodes an approximate representation of an unbounded set of elements with a bounded amount of state. It is widely used in transactional memory systems to uncover conflicts [18]. A data dependence is similar to a conflict in transactional memory because it exists only if two or more memory operations access the same memory location in some order. Therefore, a signature is also suitable for detecting data dependences.

We evaluated our approach using the NAS parallel benchmark suite (NAS) [19] and Starbench parallel benchmark suite (Starbench) [20]. The performance results are shown in **Table 1**.

Benchmark	Average slowdown			Average memory consumption (MB)	
	1T	8T	16T	8T	16T
NAS	190	97	78	473	649
Starbench	191	101	93	505	1390

Table 1. Performance results of profiler in DiscoPoP.

While performing an exhaustive dependence search with 16 profiling threads, our lock-free parallel design limited the average slowdown to 78× and 93× for NAS and Starbench, respectively. Using a signature with 10^8 slots, the memory consumption did not exceed 649 MB (NAS) and 1390 MB (Starbench).

3.2. Decomposition

The most difficult and challenging part in parallelizing sequential programs is to identify which code sections are able to run in parallel. While identifying such code sections, most of the current parallelism discovery techniques focus on specific language constructs. In contrast, we propose the concept of computational units (CUs) to concentrate on the computations performed by a program independently of any language constructs.

In our approach, a program is treated as a collection of computations communicating with one another using a number of variables. Each computation is represented as a *computational unit* (CU). A CU is a collection of instructions following the *read-compute-write* pattern: a set of variables is read and used to perform a computation, and then the result is written back to another set of variables. The two sets of variables are called *read set* and *write set*, respectively. These two sets do not necessarily have to be disjoint. Load instructions reading variables in the read set form the *read phase* of the CU, and store instructions writing variables in the write set form the *write phase* of the CU. A CU is defined by a read-compute-write pattern because, in practice, tasks communicate with one another by reading and writing variables that are global to them, while computations are performed locally.

We build a *CU graph*, in which vertices are statically generated CUs and edges are dynamic data dependences. Data dependences in a CU graph are always among instructions in read phases and write phases. Dependences that are local to a CU are hidden because they do not prevent parallelism among CUs. Our tool also generates the program execution tree (PET) of a program. This tree contains information about program control dependences and execution paths. Nodes of the tree are control regions of the program. We map the CU graph of a program onto its execution tree to determine CUs for every region. **Figure 2** shows the CU graph of a program mapped onto its PET. PET and CU graphs serve for different kinds of code analyses as they contain the information such as CUs and their corresponding instructions, data and control dependences, etc. In this work we mainly use them for parallelization.

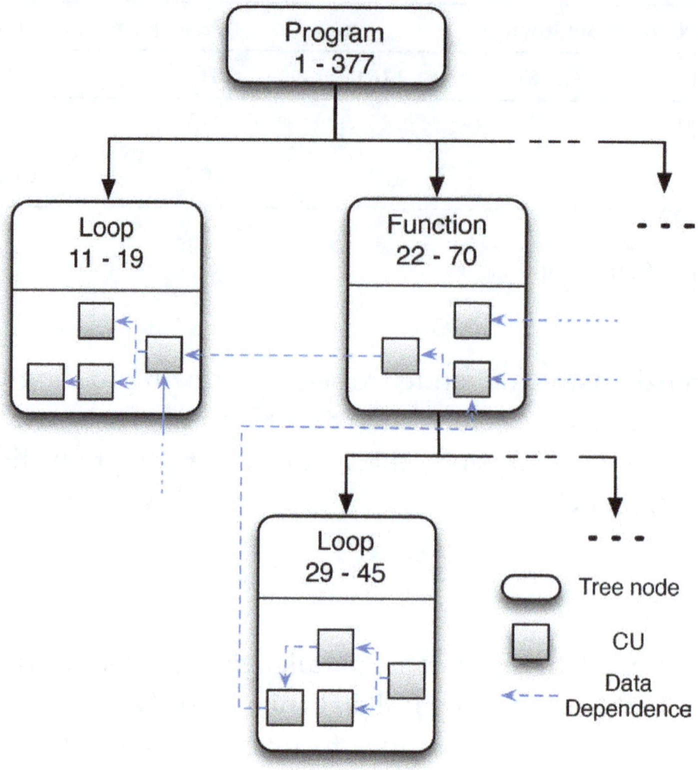

Figure 2. CU graph mapped onto the program execution tree.

4. Task extraction and parallel pattern identification

In the following section, we focus on using PET and CU graphs for parallelism discovery and supporting the parallelization process. Additionally, we describe how to use them to detect parallel design patterns in the given sequential code automatically.

4.1. Task extraction

DiscoPoP suggests parallelism among strongly connected components (SCCs) and chains in the CU graph of a program [21]. An SCC is a subgraph of the CU graph in which every CU is reachable from every other CU. It forms a complex knot of dependences that defy internal parallelization. A chain is a group of CUs that are connected in a row without a branching or joining point in between. We merge chains because a CU contains only a few instructions and there is no benefit in considering each CU as a separate task. The CUs grouped as SCCs or chains could form separate tasks and be executed in parallel, if there are no dependences between them. Parallelism is also possible when dependences are weakly connected. DiscoPoP discovers these parallelization opportunities by calculating affinity between CUs and applying the minimum cut algorithm [22] to the CU graph. It calculates the affinity for every pair of CUs based on the number of dependences and shared instructions between them. DiscoPoP suggests to partition the CU graph with a minimum number of dependences and affected shared instructions.

Finally DiscoPoP ranks parallelization opportunities to prioritize them based on three metrics [21]. The first is instruction coverage. It provides an estimate of how much time is spent in a code section. The second is speedup, which reflects potential speedup if the code section is parallelized. The third is CU imbalance. It reflects when suggested parallelization may lead to a bottleneck. Our experiments with Barcelona OpenMP Tasks Suite (BOTS) [23], NAS, PARSEC benchmark suite (PARSEC) [24], and Starbench showed that all of the code sections identified as parallelizable by our approach are parallelized in the existing parallel versions of the benchmark programs [3, 21, 25, 26].

4.2. Parallel pattern identification

Parallel design patterns are reusable solutions for common problems that occur during the development of parallel programs. They have been developed to help programmers to design and implement parallel applications efficiently [27, 28]. However, identifying a suitable parallel pattern for a specific code region in a sequential application is a difficult task. Also, transforming the application according to structures supporting those parallel patterns is very challenging.

We propose an approach that automatically finds parallel patterns in the algorithm structure design space of sequential applications using template matching [7, 8]. The approach generates a pattern vector, which plays the role of the template to be matched to the program. For each hotspot in the program, we create the pattern-specific graph vector according to the dependences of the corresponding CU graph. The correlation coefficient of the pattern vector and the graph vector of the selected hotspot tells us whether the pattern exists in the selected section or not. So far, we support the detection of pipeline, do-all, task-level parallelism such as master/worker, geometric decomposition, and reduction patterns. Our tool not only indicates whether a parallel design pattern has been found in some section of the program but also shows how the code must be divided to fit the appropriate structure of the pattern.

We evaluated our approach with 17 sequential applications from four different benchmark suites, i.e., Starbench, BOTS, PARSEC, and PolyBench [29]. We successfully detected pipeline, task parallelism, geometric decomposition, fusion, and reduction. We compared the detected patterns with the existing parallel versions of the benchmarks and confirmed our results [7, 8]. For those benchmarks for which the parallel version does not exist, we implemented the detected patterns. We achieved a speedup of 14× with 32 threads for the best case of our hand-implemented parallel version of the *ludcmp* application in PolyBench [8].

5. Code transformation

After finding parallelization opportunities in the program, generating parallel code to run on the hardware is another main step of the parallelization process. The code transformation component [9] in DiscoPoP transforms sequential C/C++ code into parallel code, following the detected parallel design pattern. Transformation is performed on the AST level using the Clang libraries [30]. The transformation module traverses the Clang AST of the source code to locate simple detected patterns such as do-all or task-level parallelism and the corresponding code sections. Then a source-code rewriting module rewrites the targeted source-code strings in the Clang AST context using Intel TBB parallel constructs—the *parallel_for* and *flow graph* templates. The transformation process does not require users to annotate parallel code sections in advance. The parallel for-loop transformation is automatic, and the parallel-task transformation using the flow graph template requires user assistance. The user needs to specify the buffering policy in the synchronization join node of the flow graph.

We have evaluated applications from NAS, PARSEC, and Intel CnC samples. The obtained results confirm that our approach is able to achieve promising performance with minor user interference. The average speedups of loop parallelization and task parallelization are up to 3.12× and 9.92×, respectively [9].

6. Correctness analysis

The parallelized code is expected to deliver the same output as its sequential counterpart. To assure the correctness, we use automated unit testing [10, 31, 32] and sophisticated race detection methods [11, 33]. For data races, our library-independent race detection approach [33] is applied to the generated parallel code for finding potential data races. Also, automated unit tests based on dynamic and static analyses are generated, which can be used during the parallelization process for finding atomicity violations and verifying the parallel code.

A notorious class of concurrency bugs is race conditions related to nonatomic updates on correlated variables, potentially leading to broken invariants, which make up about 30% of all non-deadlock concurrency bugs. We propose to combine the benefits of automatic parallel unit test generation with the advantages of race detection. To achieve this, our framework uses the existing unit test generator AutoRT [34, 35] to identify possible correlation violations in function pairs accessing correlated variables. We automatically generated 81 parallel unit tests for correlated variables in eight different applications. After analyzing the unit tests, a race detector for correlated variables reported more than 85% of the race conditions violating variable correlations [36]. Furthermore, we were able to reduce the number of redundantly generated tests by up to 50%.

7. Further applications of the DiscoPoP framework

Considering the modern parallel programming models and hardware platforms, communication patterns play an important role for energy efficiency and performance of the generated parallel code. We investigated communication patterns in shared-memory applications, which are useful for applying optimizations and finding performance bottlenecks.

7.1. Communication pattern detection for auto-tuning

Communication patterns extracted from parallel programs can provide a valuable source of information for auto-tuning and runtime workload scheduling on heterogeneous systems. Once identified, such patterns can help find the most promising optimizations. Communication patterns can be detected using different methods, including sandbox simulation, memory profiling, and hardware counter analysis. However, these analyses usually suffer from high runtime and memory overhead, necessitating a tradeoff between accuracy and resource consumption. More importantly, none of the existing methods exploit fine-grained communication patterns on the level of individual code regions.

We extended the DiscoPoP profiler by adding a communication pattern detection component and employed an asymmetric signature memory method to detect communication among threads [12]. Shared-memory systems have fundamental differences in comparison with distributed memory system. Shared-memory applications bring additional irregularity and complexity to data sharing, which imposes further difficulty on finding the communication pattern. Communications are implicit and automatically occur through memory accesses, when one thread writes a value and another one reads it. We experimentally validated our communication pattern detection approach with programs in the SPLASH [37] benchmark suite and successfully identified the typical communication patterns existing in parallel programs [12]. The runtime overhead of our extended profiler is around 225×, while the required amount of memory remains fixed.

7.2. Optimization techniques for transactional memory

Transactional memory (TM) is a promising paradigm that facilitates programming for shared-memory systems. We used DiscoPoP and reported optimization techniques in software transactional memory (STM) [13]. We demonstrated that varying STM parameters such as the size of transaction, readset, and writeset significantly change the execution time of the STM programs. By applying machine learning and using DiscoPoP results, we optimized these parameters. The experimental results with NAS revealed that we are able to improve the performance of STM programs by up to 54.8%. In another work [14], we used DiscoPoP for restricted transactional memory (RTM) on Intel's Haswell processor and showed that the performance of RTM varies across applications. While RTM enhances performance of some applications relative to software transactional memory (STM), it degrades performance in some others. Using DiscoPoP, we proposed an adaptive system that switches between HTM and STM in transaction granularity and predicts the optimal TM system for a given transaction.

8. Limitations

Methods to analyze programs are generally divided into two categories: static and dynamic methods. Static methods analyze source or intermediate code and are restricted to information that can be obtained before running the program, i.e., at compile time. Static approaches are fast but also conservative because they have limited support for runtime information. In contrast, dynamic approaches identify dependences only if they exist at runtime. Although dynamic approaches relax the conservative assumptions made by static approaches on dynamic data, they are input sensitive, that is, their outcome may depend on the particular program execution. To mitigate this limitation, dynamic approaches generally execute the target program with a range of representative inputs, a practice we adopt as well when using our dynamic dependence profiler. Additionally, we plan to derive conditional correctness guarantees, taking the specific nature of the missing dependences into account. This would allow users to run the program in parallel if the missed dependences are irrelevant for a given input configuration.

Our parallelism detection approach is dependent on the profiler's output. Due to the use of signature technology (as an approximation), the dependence profiler could have a very low rate of false positives and false negatives. If these appear in the profiler results, then our dependence analysis and accordingly the parallelism detection results could be affected.

The pattern detection approach is dependent on the coding style of a programmer. For example in a Starbench program, we found a pipeline pattern because the programmer used pointers to access arrays and used the increment operator (++) on pointers. However, if a loop index variable (loop indexing) had been used in the sequential code, we would have detected a do-all loop pattern based on our pattern detection algorithm (template matching).

MPMD-style task parallelism (Multiple Program Multiple Data) could be found in few evaluated benchmarks, which lead to minor speedups. More intensive investigations are needed to apply such kind of parallelism to real-world applications.

Recently, we developed a prototypical visualization component to display the output of DiscoPoP. However, a more advanced and user-friendly graphical user interface to visually guide application developers in a stepwise manner when parallelizing a program would be desirable. This feature could create higher incentives for developers and make the parallelization workflow easier, particularly for DiscoPoP's code-based semiautomatic parallelization approach.

9. Related work

Profiling and parallelism discovery has always been a central topic in the field of parallel programming. Early approaches mainly analyze source code statically and predict parallelism based on theoretical models [38, 39]. Bobbie [40] presented a method to partition a program for parallelization. The method adopts syntax-driven data-dependence analysis and detects parallelism based on Bernstein's conditions [41]. It uses bipartite graph matching to partition the code.

Compiler-based auto-parallelization based on the polyhedral model [1, 2, 42] is generally restricted to program loops with specific criteria (e.g., *affine* linear loop boundaries and array indices). A dynamically speculative extension of these criteria expands the applicability of this method to a certain extent [43]. Another work [28] that is not restricted to loops only identifies parallel tasks in the static dependence graphs using integer linear programming. Generally, compiler-based auto-parallelization is often conservative and fails to identify available parallelism for many applications, because runtime information such as the values of pointers and array indices are often not known at compile time. In practice, the parallelization of software usually happens manually, and often, it is more appropriate to follow the provided guidelines for parallel design patterns [44].

Without considering runtime behavior of the target program, some key parallelism usually could remain undetected in static approaches. To overcome this disadvantage, profiling techniques to gather runtime data emerged [45, 46]. Such methods are usually referred to as dynamic parallelism discovery approaches. Additionally, most of these approaches have a cost model to produce results. Kremlin [47] determines the length of the critical path in a given code region. Based on this knowledge, Kremlin calculates a metric called self-parallelism to quantify the parallelism of a code region. The tool reports self-parallelism for each region in a descending order. Alchemist [48] identifies predefined constructs that can be treated as candidates for asynchronous execution in sequential programs. It estimates the effectiveness of parallelizing a certain construct using Valgrind [49]. Kremlin and Alchemist mainly focus on loops, which are easier to profile and quantify.

At the same time, other dynamic parallelism discovery approaches deal with task parallelism as tasking became popular and widely supported in almost all mainstream parallel programming libraries and frameworks. Ketterlin et al. [50] profiles sequential programs and represents them using execution trees. It further attaches data dependences to the nodes of the execution tree and discovers task parallelism where two or more nodes are independent of one another. The SLX Tool Suite, formerly known as MAPS [51], concentrates on parallelism discovery for applications on multiprocessor system on chip (MPSoC). It identifies code sections called *coupled blocks*. These code blocks are identified with constraints requiring that they should be schedulable and should be tightly coupled by data dependences. Each coupled block is considered as a task, and two tasks can run in parallel if there is no data dependence between them.

Tareador [52] provides a set of annotations for marking down tasks in the code. It takes a relatively brute-force approach by enumerating possible decompositions and does not take the control flow into account, which may lead to tasks that are not easy to implement. Intel Advisor XE [53] is a prototyping tool for different programming languages such as C, C++, C#, and Fortran. It also performs a correctness check, which is essentially a data-race detector and has a large time overhead. Also pattern detection and code transformation are not supported by Intel Advisor XE.

The approach presented by Tournavitis et al. [54] uses both static analysis and dynamic profiling to detect potential parallelism. A machine learning-based prediction mechanism maps the parallelism onto different architectures. It generates parallel code using OpenMP annotations and targets loop-based parallelism. However, the code transformation is relatively simple. The tool does not perform high-level code restructuring that could exploit coarse-grained task parallelism. In recent work [55], the tool exploits pipeline parallelism.

OpenRefactoryC [56] is a tool providing many refactoring methods for C programs, but it does not automatically transform sequential code to parallel code. The approach presented in [57] transforms serial C++ code to parallel code using OpenMP directives. However, it requires users to define the high-level abstractions in advance.

Similar to all the dynamic parallelism discovery approaches, the DiscoPoP approach adopts profiling techniques to gather runtime data. However, our method discovers parallelism based on computational units (CUs), which are derived statically, and parallel design patterns. We identify CUs in sequential programs and build the CU graph as the representation of a program. Based on the CU graph, we can perform different analyses and detect parallel design patterns. A CU clearly distinguishes the inputs and outputs of a computation, allowing a direct application of Bernstein's conditions [41]. Bernstein's conditions describe when two program segments are independent and can be executed in parallel. In addition, our method discovers both task-based and loop-based parallelisms using the same framework. A CU in our approach acts as a task, a stage in a pipeline, or an iteration of a loop or a subset of either of these based on the context, which distinguishes our work from related work.

10. Conclusion and outlook

In this chapter, we propose an optimistic code-based approach to assist semiautomatic parallelization in multicore architectures, focusing on general-purpose applications. Our approach is implemented as an integrated tool based on LLVM. Program analysis and parallelism discovery are performed at the LLVM-IR level and are not limited to any programming language or specific language constructs.

The proposed approach presents an alternative to conservative and usually loop-centric auto-parallelization. Application developers can take advantage of our methods to identify and exploit parallelism that is not necessarily limited to loop parallelism for many applications. Our semiautomatic approach can reduce the high cost and price of the manual error-prone parallelization process. At the same time, many legacy applications can benefit from the available hardware parallelism using our approach.

There is no doubt that parallel programming is challenging and involves a steep learning curve. Developers must think about the application in new ways. It is possible to work months on parallelizing an application and end up with incorrect results, or the resulting parallel program runs slower than the sequential one. For this reason, the techniques and tools used for parallelism discovery, debugging, and tuning the performance during the parallelization process play a very significant role. Our approach supports application developers during this process. It encourages average programmers to use parallel programming by creating incentives and insight for developers and making the parallelization workflow easier.

Considering the hardware trend and future smart cyber-physical systems (smart factory 4.0), energy-efficient programming is a key feature in improving productivity and efficiency.

Whether we develop an application for mobile devices or data centers, we want to reduce energy, e.g., to increase the battery life of a mobile device or lower the customer's data center utility bill. Thus, the role of programmers to reduce the energy and develop power-efficient applications is very important. Our future work considers energy-efficient software development during the parallelization process. Ongoing work focuses on developing an energy efficiency method to be integrated in our parallelization approach. Energy conservation without performance degradation is challenging and has become an important trend. Our initial results suggest that we can propose energy-efficient task decomposition and programming constructs during the parallelization process. Our preliminary evaluation shows up to 21% improvements of energy consumption after applying our optimizations. Our overarching goal is to improve efficiency while maintaining productivity.

Acknowledgements

I would like to thank all my colleagues and collaborators who contributed to the results described in this chapter. In particular, I would like to thank Ehsan Atoofian (Lakehead University, Canada), Rohit Atre (TU Darmstadt, Germany), Michael Beaumont (RWTH Aachen University, Germany), Daniel Fried (UC Berkeley, USA), Michael Gerndt (TU Munich, Germany), Wolfram Gottschlich (University of Passau, Germany), Kurt Keutzer (UC Berkeley, USA), Nico Koprowski (Daimler AG, Germany), Zhen Li (SAP, Germany), Arya Mazaheri (TU Darmstadt, Germany), Korbinian Molitorisz (Agilent Technologies, Germany), Mohammad Norouzi (TU Darmstadt, Germany), Jochen Schimmel (Karlsruhe Institute of Technology, Germany), Thireshan Jeyakumaran (Lakehead University, Canada), Walter Tichy (Karlsruhe Institute of Technology, Germany), Zia Ul Huda (TU Darmstadt, Germany), Felix Wolf (TU Darmstadt, Germany), Yang Xiao (Lakehead University, Canada), and Bo Zhao (Humboldt University of Berlin).

Author details

Ali Jannesari

Address all correspondence to: jannesari@iastate.edu

Department of Computer Science, Iowa State University, USA

References

[1] Feautrier P. Automatic parallelization in the polytope model. In: The Data Parallel Programming Model: Foundations, HPF Realization, and Scientific Applications. London: Springer-Verlag; 1996. pp. 79-103. [Online]. Available: http://dl.acm.org/citation.cfm?id = 647429.723579

[2] Griebl M, Lengauer C, Wetzel S. Code generation in the polytope model. In: Proceedings of the 1998 International Conference on Parallel Architectures and Compilation Techniques, ser. PACT '98. Washington, DC: IEEE Computer Society; 1998. p. 106. [Online]. Available: http://dl.acm.org/citation.cfm?id=522344.825673

[3] Li Z, Atre R, Huda ZU, Jannesari A, Wolf F. Unveiling parallelization opportunities in sequential programs. Journal of Systems and Software. July 2016;**117**:282-295

[4] Li Z, Jannesari A, Wolf F. An efficient data-dependence profiler for sequential and parallel programs. In: Proceedings of the 29th IEEE International Parallel and Distributed Processing Symposium (IPDPS). Hyderabad, India: IEEE Computer Society; May 2015. pp. 484-493

[5] Atre R, Jannesari A, Wolf F. The basic building blocks of parallel tasks. In: Proceedings of the International Workshop on Code Optimisation for Multi and Many Cores; San Francisco, CA. ACM; February 2015. pp. 1-12

[6] Li Z, Jannesari A, Wolf F. Discovery of potential parallelism in sequential programs. In: Proceedings of the 42nd International Conference on Parallel Processing Workshops (ICPPW), Workshop on Parallel Software Tools and Tool Infrastructures (PSTI); Lyon, France. October 2013. pp. 1004-1013

[7] Huda ZU, Jannesari A, Wolf F. Using template matching to infer parallel design patterns. ACM Transactions on Architecture and Code Optimization. 21 January 2015;**11**(4):1-64

[8] Huda ZU, Atre R, Jannesari A, Wolf F. Automatic parallel pattern detection in the algorithm structure design space. In Proceedings of the 30th IEEE International Parallel and Distributed Processing Symposium (IPDPS); Chicago. IEEE Computer Society; May 2016. pp. 43-52

[9] Zhao B, Li Z, Jannesari A, Wolf F, Wu W. Dependence-based code transformation for coarse-grained parallelism. In: Proceedings of the International Workshop on Code Optimisation for Multi and Many Cores; San Francisco. ACM; February 2015. pp. 1-10

[10] Jannesari A, Wolf F. Automatic generation of unit tests for correlated variables in parallel programs. International Journal of Parallel Programming (IJPP). March 2016;**44**(3):644-662 [Online]. Available: http://dx.doi.org/10.1007/s10766-015-0363-8

[11] Jannesari A. Detection of high-level synchronization anomalies in parallel programs. International Journal of Parallel Programming (IJPP). August 2015;**43**(4):656-678

[12] Mazaheri A, Jannesari A, Mirzaei A, Wolf F. Characterizing loop-level communication patterns in shared memory applications. In: Proceedings of the 44th International Conference on Parallel Processing (ICPP); Beijing, China. September 2015. pp. 759-768

[13] Xiao Y, Li Z, Atoofian E, Jannesari A. Automatic optimization of software transactional memory through linear regression and decision tree. In: Proceedings of 15th International Conference on Algorithms and Architectures for Parallel Processing (ICA3PP); Zhangjiajie, China, ser. Lecture Notes in Computer Science, Vol. 9531. Springer International Publishing; November 2015. pp. 61-73

[14] Jeyakumaran T, Atoofian E, Xiao Y, Li Z, Jannesari A. Improving performance of transactional applications through adaptive transactional memory. In: Proceedings of the 24th Euromicro International Conference on Parallel, Distributed, and Network-Based Processing (PDP); Heraklion Crete, Greece. February 2016

[15] Rul S, Vandierendonck H, De Bosschere K. A profile-based tool for finding pipeline parallelism in sequential programs. Parallel Computing. September 2010;**36**(9):531-551

[16] Kim M, Kim H, Luk CK. SD3: A scalable approach to dynamic data- dependence profiling. In: Proceedings of the 43rd Annual IEEE/ACM International Symposium on Microarchitecture, ser. MICRO 43. IEEE Computer Society; 2010. pp. 535-546

[17] Fraser K, Harris T. Concurrent programming without locks. ACM Transactions on Computer System. May 2007;**25**(2). DOI: 10.1145/1233307.1233309

[18] Sanchez D, Yen L, Hill MD, Sankaralingam K. Implementing signatures for transactional memory. In: Proceedings of the 40th Annual IEEE/ACM International Symposium on Microarchitecture, ser. MICRO 40. IEEE Computer Society; 2007. pp. 123-133

[19] Bailey DH, Barszcz E, Barton JT, Browning DS, Carter RL, Fa-toohi RA, Frederickson PO, Lasinski TA, Simon HD, Venkatakrishnan V, Weeratunga SK. The NAS parallel benchmarks. The International Journal of Supercomputer Applications. 1991;**5**(3):63-73

[20] Andersch M, Juurlink B, Chi CC. A benchmark suite for evaluating parallel programming models. In: Proceedings of the 24th Workshop on Parallel Systems and, Algorithms, ser. PARS '11. 2011. pp. 7-17

[21] Li Z, Jannesari A, Wolf F. Discovering parallelization opportunities in sequential programs—A closer-to-complete solution. In: Proceedings of the First International Workshop on Software Engineering for Parallel Systems. 2014. pp. 1-10

[22] Luxburg U. A tutorial on spectral clustering. Statistics and Computing. December 2007;**17**(4):395-416 [Online]. Available: http://dx.doi.org/10.1007/ s11222-007-9033-z

[23] Duran A, Teruel X, Ferrer R, Martorell X, Ayguade E. Barcelona openmp tasks suite: A set of benchmarks targeting the exploitation of task parallelism in openmp. In: Proceedings of the 2009 International Conference on Parallel Processing, ser. ICPP '09. Washington, DC: IEEE Computer Society; 2009. pp. 124-131. [Online]. Available: http://dx.doi.org/10.1109/ICPP.2009.64

[24] Bienia C, Kumar S, Singh JP, Li K. The parsec benchmark suite: Characterization and architectural implications. In: Proceedings of the 17th International Conference on Parallel Architectures and Compilation Techniques, ser. PACT '08. New York: ACM; 2008. pp. 72-81. [Online]. Available: http://doi.acm.org/10.1145/1454115.1454128

[25] Li Z, Atre R, Ul-Huda Z, Jannesari A, Wolf F. Discopop: A profiling tool to identify parallelization opportunities. In: Tools for High Performance Computing 2014. Springer International Publishing; August 2015, ch. 3. pp. 37-54

[26] Li Z, Zhao B, Jannesari A, Wolf F. Beyond data parallelism: Identifying parallel tasks in sequential programs. In: Proceedings of 15th International Conference on Algorithms and Architectures for Parallel Processing (ICA3PP); Zhangjiajie, China, ser. Lecture Notes in Computer Science, Vol. 9531. Springer International Publishing, November 2015. pp. 569-582

[27] Jahr R, Gerdes M, Ungerer T. A pattern-supported parallelization approach. In: Proceedings of the 2013 International Workshop on Programming Models and Applications for Multicores and Manycores, ser. PMAM '13. New York: ACM; 2013. pp. 53-62. [Online]. Available: http: //doi.acm.org/10.1145/2442992.2442998

[28] Streit K, Doerfert J, Hammacher C, Zeller A, Hack S. Generalized task parallelism. ACM Transactions on Architecture and Code Optimization. April 25, 2015;**12**(1):1-8. [Online]. Available: http://doi.acm.org/10.1145/2723164

[29] http://www.cs.ucla.edu/~pouchet/software/PolyBench/.

[30] http://clang.llvm.org.

[31] Jannesari A, Koprowski N, Schimmel J, Wolf F. Generating classified parallel unit tests. In: Tests and Proofs: Proceedings of the 8th International Conference, TAP 2014, Held as Part of STAF 2014; July 24-25, 2014; York. Springer International Publishing; December 2014. pp. 117-133

[32] Jannesari A, Koprowski N, Schimmel J, Wolf F, Tichy WF. Detecting correlation violations and data races by inferring non-deterministic reads. In: 2013 International Conference on Parallel and Distributed Systems (ICPADS). December 2013. pp. 1-9

[33] Jannesari A, Tichy WF. Library-independent data race detection. IEEE Transactions on Parallel and Distributed Systems (TPDS). 2013;**PP**(99):1-11

[34] Schimmel J, Molitorisz K, Jannesari A, Tichy WF. Automatic generation of parallel unit tests. In: Proceedings of the 8th International Workshop on Automation of Software Test (AST); San Francisco. ACM; May 2013. pp. 40-46

[35] Schimmel J, Molitorisz K, Jannesari A, Tichy WF. Combining unit tests for data race detection. In: Proceedings of 10th IEEE/ACM International Workshop on Automation of Software Test (AST 2015). IEEE; May 2015. pp. 43-47. [Online]. Available: http://dl.acm.org/citation.cfm?id=2819261. 2819275

[36] Jannesari A, Westphal-Furuya M, Tichy WF. Dynamic data race detection for correlated variables. In: Proceedings of the 11th International Conference on Algorithms and architectures for parallel processing—Volume Part I, ser. ICA3PP'11. Berlin, Germany: Springer-Verlag; 2011. pp. 14-26. [Online]. Available: http://dl.acm.org/citation.cfm?id=2075416.2075421

[37] Woo SC, Ohara M, Torrie E, Singh JP, Gupta A. The splash-2 programs: Characterization and methodological considerations. In: Proceedings of the 22Nd Annual International Symposium on Computer Architecture, ser. ISCA '95. New York: ACM; 1995. pp. 24-36. [Online]. Available: http://doi.acm.org/10.1145/223982.223990

[38] Burke M, Cytron R, Ferrante J, Hsieh W. Automatic generation of nested, fork-join paral

lelism. The Journal of Supercomputing. 1989;**3**(2):71-88 [Online]. Available: http://dx.doi.org/10.1007/BF00129843

[39] Sarkar V. Automatic partitioning of a program dependence graph into parallel tasks. IBM Journal of Research and Development. 1991;**35**(5.6):779-804

[40] Bobbie P. Partitioning programs for parallel execution: A case study in the Intel iPSC/2 environment. International Journal of Mini & Microcomputers. 1997;**19**(2):84-96

[41] Bernstein A. Analysis of programs for parallel processing. IEEE Transactions on Electronic Computers. 1966;**15**(5):757-763

[42] Bondhugula U, Hartono A, Ramanujam J, Sadayappan P. A practical automatic polyhedral parallelizer and locality optimizer. In: Proceedings of the 29th ACM SIGPLAN Conference on Programming Language Design and Implementation, ser. PLDI '08. New York: ACM; 2008. pp. 101-113. [Online]. Available: http://doi.acm.org/10.1145/1375581.1375595

[43] Martinez Caamano JM, Wolff W, Clauss P. Code Bones: Fast and Flexible Code Generation for Dynamic and Speculative Polyhedral Optimization. Cham: Springer International Publishing; 2016. pp. 225-237 [Online]. Available: http://dx.doi.org/10.1007/978-3-319-43659-3_17

[44] Mattson T, Sanders B, Massingill B. Patterns for Parallel Programming. 1st ed. Boston: Addison-Wesley Professional; 2004

[45] Rul S, Vandierendonck H, De Bosschere K. A profile-based tool for finding pipeline parallelism in sequential programs. Parallel Computing. 2010;**36**(9):531-551

[46] Huang J, Jablin TB, Beard SR, Johnson NP, August DI. Automatically exploiting cross-invocation parallelism using runtime information. In: IEEE/ACM International Symposium on Code Generation and Optimization (CGO). IEEE; 2013. 1-11

[47] Garcia S, Jeon D, Louie CM, Taylor MB. Kremlin: Rethinking and rebooting gprof for the multicore age. SIGPLAN Notices. June 2011;**46**(6):458-469. [Online]. Available: http://doi.acm.org/10.1145/1993316.1993553

[48] Zhang X, Navabi A, Jagannathan S. Alchemist: A transparent dependence distance profiling infrastructure. In: Proceedings of the 7th annual IEEE/ACM International Symposium on Code Generation and Optimization. IEEE Computer Society; 2009. 47-58

[49] Nethercote N, Seward J. Valgrind: A framework for heavyweight dynamic binary instrumentation. SIGPLAN Notices. June 2007;**42**(6):89-100 [Online]. Available: http://doi.acm.org/10.1145/1273442.1250746

[50] Ketterlin A, Clauss P. Profiling data-dependence to assist parallelization: Framework, scope, and optimization. In: Proceedings of the 2012 45th Annual IEEE/ACM International Symposium on Microarchitecture. IEEE Computer Society; 2012. 437-448

[51] Ceng J, Castrillon J, Sheng W, Scharwächter H, Leupers R., Ascheid G, Meyr H, Isshiki T, Kunieda H. Maps: An integrated framework for mpsoc application parallelization. In:

Proceedings of the 45th Annual Design Automation Conference, ser. DAC '08. ACM; 2008. pp. 754-759

[52] Subotic V, Ayguadé E, Labarta J, Valero M. Automatic Exploration of Potential Parallelism in Sequential Applications. Cham: Springer International Publishing; 2014. pp. 156-171 [Online]. Available: http://dx.doi.org/10.1007/ 978-3-319-07518-1_10

[53] http://software.intel.com/en-us/intel-advisor-xe.

[54] Tournavitis G, Wang Z, Franke B, O'Boyle MF. Towards a holistic approach to auto-parallelization: Integrating profile-driven parallelism detection and machine-learning based mapping. In: Proceedings of the 2009 ACM SIGPLAN Conference on Programming Language Design and, Implementation, ser. PLDI '09. ACM; 2009. pp. 177-187. [Online]. Available: http://doi.acm.org/10.1145/1542476.1542496

[55] Tournavitis G, Franke B. Semi-automatic extraction and exploitation of hierarchical pipeline parallelism using profiling information. In: Proceedings of the 19th International Conference on Parallel Architectures and Compilation Techniques, ser. PACT '10. ACM; 2010. 377-388. [Online]. Available: http://doi.acm.org/10.1145/1854273.1854321

[56] Hafiz M, Overbey J, Behrang F, Hall J. Openrefactory/c: An infrastructure for building correct and complex c transformations. In: Proceedings of the 2013 ACM Workshop on Refactoring Tools, ser. WRT '13. ACM; 2013. 1-4. [Online]. Available: http://doi.acm.org/10.1145/2541348.2541349

[57] Quinlan D, Schordan M, Yi Q, de Supinski BR. A c++ infrastructure for automatic introduction and translation of openmp directives. In: OpenMP Shared Memory Parallel Programming. Springer; 2003. pp. 13-25. [Online]. Available: http://dx.doi.org/10.1007/3-540-45009-2_2

7

Energy Savings in EAF Steelmaking by Process Simulation and Data-Science Modeling on the Reproduced Results

Panagiotis Sismanis

Abstract

Electric-Arc-Furnace (EAF)-based process route in modern steelmaking for the production of plates and special quality bars requires a series of stations for the secondary metallurgy treatment (Ladle-Furnace, and potentially Vacuum-Degasser), till the final casting for the production of slabs and blooms in the corresponding continuous casting machines. However, since every steel grade has its own melting characteristics, the melting (liquidus) temperature per grade is generally different and plays an important role in the final casting temperature, which has to exceed by somewhat the melting temperature by an amount called superheat. The superheat is adjusted at the ladle-furnace (LF) station by the operator who decides mostly on personal experience but, since the ladle has to pass from downstream processes, the liquid steel loses temperature not only due to the duration of the processes till casting but also due to the ladle refractory history. Simulation software was developed in order to reproduce the phenomena involved in a meltshop and influence downstream superheats. Data science models were deployed in order to check the potential of controlling casting temperatures by adjusting liquid-steel exit temperatures at LF.

Keywords: continuous casting, superheat, billet, slab, grade, supervised model, simulation

1. Introduction

The effect of superheat (SPH) on the potential of surface and sub-surface defects generation in the continuous cast products is known for many years. Ayata et al. [1] have pointed out the advantage of low SPH teeming upon product quality since 1995, and in the same year Thomas [2] has discussed the need to include SPH in thermal-mechanical models for continuous casting. Guyot et al. [3] have discussed the effect of SPH on surface quality issues in peritectic slabs. Jansto [4] has pointed out the effect of SPH on quality issues for Nb-based microalloyed steels. Jacobi and Schwerdtfeger [5] talking about ripple marks on cast steel surfaces have notified the importance of keeping SPH values low at casting; furthermore, if the superheat is too high for a grade, it may give rise to defects in the product. As 10°C increase per ton of liquid steel requires theoretically 2.2 kWh of electrical energy [6], one may realize the energy lost annually at casting if superheat is much larger than required. It is understood that a system that will notify the LF operator to adjust the liquid steel SPH in order to match the required casting temperature later on at the continuous caster is of paramount importance and is under research for a long time. Offline models based on heat transfer and thermodynamics have been developed in the past, but the focus is mostly appropriate to online statistical models which are faster to be generated and can be tuned. Nevertheless, due to the nature of liquid steel processing there is still a great deal of work on the subject to be carried out to reach this milestone. Gupta and Chandra [7] have developed a coupled heat transfer and a simple regression model in order to manage to control SPH at the caster floor; great attention was given to the holding time of liquid steel in the ladle, as well as the ladle turnaround time, that is, the time from teeming till next tapping for a ladle; a fourth-degree polynomial was derived as a regression formula simulating the initial temperature at the tundish (T_{tun1}) as a function of the holding time (t), ladle life (LL), ladle turnaround time (TAT), exit temperature at the LF (T_{LF}), and previous liquid-steel in the tundish temperature (T_{past}):

$$T_{tun1} = f + 0.019LL - 0.012TAT + 0.358T_{past} + 0.631T_{LF} \tag{1}$$

Where:

$$f = 180.912 - 40.428t + 3.173t^2 - 0.107t^3 + 0.001t^4 \tag{2}$$

Based on plant data the regression coefficient R^2 was found to be 0.73. Addes et al. [8] tried to control the casting superheat temperature by specific factors depending upon the heat sequence in the tundish, steel residence in the ladle, grade, ladle condition, tundish preheat time, and casting speed. On the other hand, Fredman et al. [9] applied the solution of the heat transfer equation in 2D in order to simulate the thermal state of the ladles. Tian et al. [10, 11] developed a hybrid model based on the energy transfer at the LF and by deploying the ensemble ELM algorithm using the modified AdaBoost.RT method to train and validate the model by plant data that were collected from a 300 t LF. Chen et al. [12, 13] have developed a model that recommends the liquid-steel exit temperature at LF in order to achieve the proper casting SPH; the model follows the input and output liquid steel energy in a ladle; it has been applied in a steelmaking plant. Sonoda et al. [14] have also developed a statistical model for predicting the liquid steel temperature at the casting floor. In the recent years, ladle-tracking systems [15, 16] have been developed that follow the route of each ladle and in this way the refractory history can be recorded; consequently, a more reliable statistical model can be developed that will predict the casting floor superheat temperatures by time. A Monte-Carlo resembling simulation software was developed for this study in order to reproduce the phenomena involved in a meltshop with respect to process times, ladle-refractory history, vacuum degasser (VD) or not treatment, and 30 different grades for blooms and slabs produced in the Stomana plant, Pernik, Bulgaria. The purpose of this study was to illustrate the potential benefits of the installation of a ladle tracking system giving online data to a supervising data-science model that will ultimately notify the proper superheat adjustment to the LF operator. On this basis, two data-science models (a distributed random forest, DRF, and a gradient boosting machine, GBM) were deduced to analyze the reproduced data. DRF and GBM models were also deduced from existing plant data and even though these data did not come from a ladle tracking system, the analysis of variance exhibited an important statistical significance. Furthermore, a GBM model was derived for the prediction of the first liquid-steel SPH at the tundish following the problem formulation of Gupta et al. [7].

2. Preparation of tests

2.1. Simulation tests

The approach to come up with a solution to the problem consisted of two procedures: at first, a Monte-Carlo type of simulation [17] was developed in order to quantify the effect of various parameters upon the required superheat (SPH) correction at the ladle-furnace (LF) station, as well as the final attained SPH at the continuous casters; second, the generated results were fed into machine-learning systems in order to identify the degree of correlation of predicted superheat values at the casting machines with respect to the reproduced corrected SPH values at the LF. **Table 1** presents the selected times for the processes involved in the computations:

Although two different casters were involved in the computations, the same transfer-time values from LF or VD were used. The simulation software was developed exclusively in R [18], as it has unique programming instructions for simulation purposes. For example, the following two commands generate 10,000 EAF process-time values derived from a normal distribution with an average value of $\mu = 60.0$ and a standard deviation value of $\sigma = 10.0$:

$$\begin{aligned} &HeatNr < -1{:}10000 \\ &EAF\_Proc < -rnorm(HeatNr, 60.0, 10.0) \end{aligned} \tag{3}$$

The greatest advantage R has is the very fast execution of instructions that are written in a form compatible for vectorization. Commands similar to (3) were written for the generation of process-time values for the rest of the processes illustrated in **Table 1**.

Twenty percent from heats produced by the EAF pass through VD treatment; furthermore, 97.5% from the VD-treated steels were selected to be billets (or blooms) and the rest slabs. The thermal history of a ladle refractory-insulation is of paramount importance for the amount of heat the contained liquid steel will absorb during reheating at the LF. Every time a ladle is placed in the position for tapping from the furnace, it may come from previous heat (almost immediately after casting) or from a refractory maintenance process that has taken some adequate time to resist the liquid-steel temperature increase at LF by absorbing some heat. The refractory insulation has also some life cycle so a new ladle may come into the production cycle at some point. **Table 2** presents some plant data related to ladle refractory maintenance that were taken under consideration in the development of the simulation program together with the need for extra liquid-steel temperature (SPH).

Process	Average (μ) process time, min	Standard deviation (σ), min
EAF	60.0	10.0
EAF to LF transfer time	10.0	2.5
LF	50.0	10.0
LF to VD transfer time	15.0	3.0
LF to CCM transfer time	12.0	3.5
VD	50.0	7.0
VD to CCM transfer time	10.0	3.5
SCCM (slab caster; CCM)	45.0	5.5
BCCM (bloom caster; CCM)	65.0	10.0

Table 1. Process standard times.

Type of maintenance or ladle condition	Average number of heats	Standard deviation	Extra SPH required	
			Average (°C)	Standard deviation
New ladle	85.0	8.0	50.0	15.0
Plates change	28.333	3.0	20.0	5.0
Inner nozzle change	10.625	1.5	30.0	8.0
Porous plug change	6.538	1.5	30.0	8.0
Slag zone repair	1.0	0.5	50.0	15.0
Immediately after previous heat	30.0	5.0	10.0	2.5
Normal preheating	8.0	6.0	10.0	3.0
Idle state	2.0	1.5	40.0	10.0

Table 2. Ladle refractory maintenance data.

Figure 1 presents the process times that were taken under consideration in the simulation part; the total process time is the sum of (1) the actual process time, that is, the total time spent at EAF, LF, VD (if the grade is VD-treated), and CCM time (which is either the bloom caster, BCCM, and slab caster, SCCM, depending upon the nature of the cast product which may be billet/bloom or slab, respectively) and (2) the transfer time, that is, the time required for the liquid steel movement between the process stations.

Liquid steel is transferred from the EAF to the LF station, then it may be transferred directly to the caster or to the VD station if this type of treatment is required, and then finally to the CCM (BCCM or SCCM). **Figure 2** depicts the time spent in this type of transfer and this is generated in the simulation software. Since VD-treated production is limited to 20% of the products, the average transfer values from LF to VD, and VD to CCM are small; on the other hand, since LF may send the ladle directly to CCM, or via VD, it is realized that two regions of points can be accumulated.

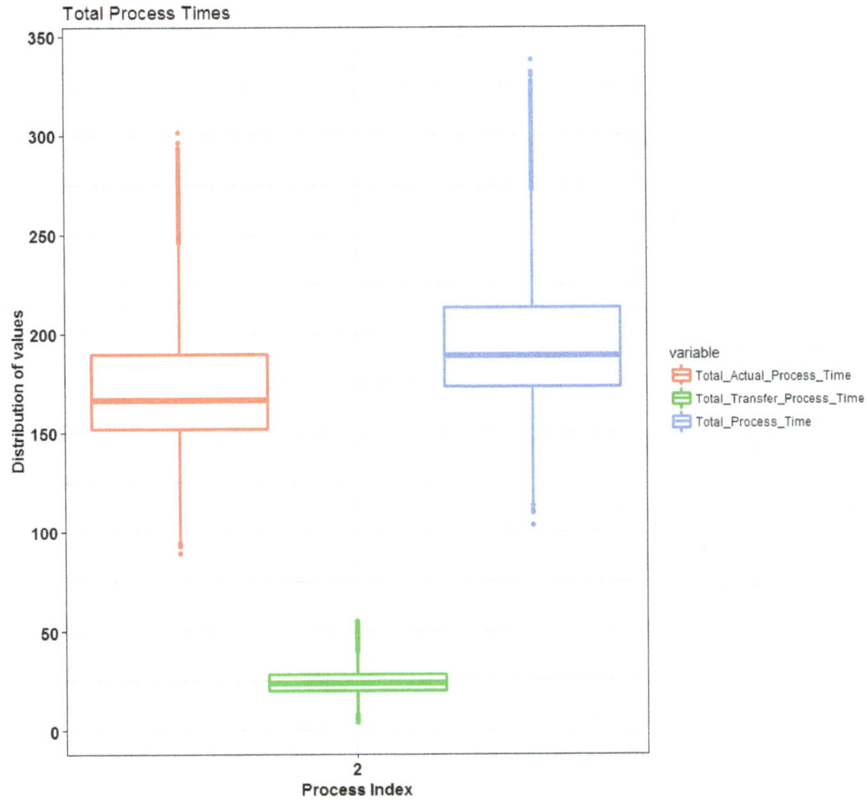

Figure 1. Boxplots representation of the distribution of the process times, actual, transfer, and total.

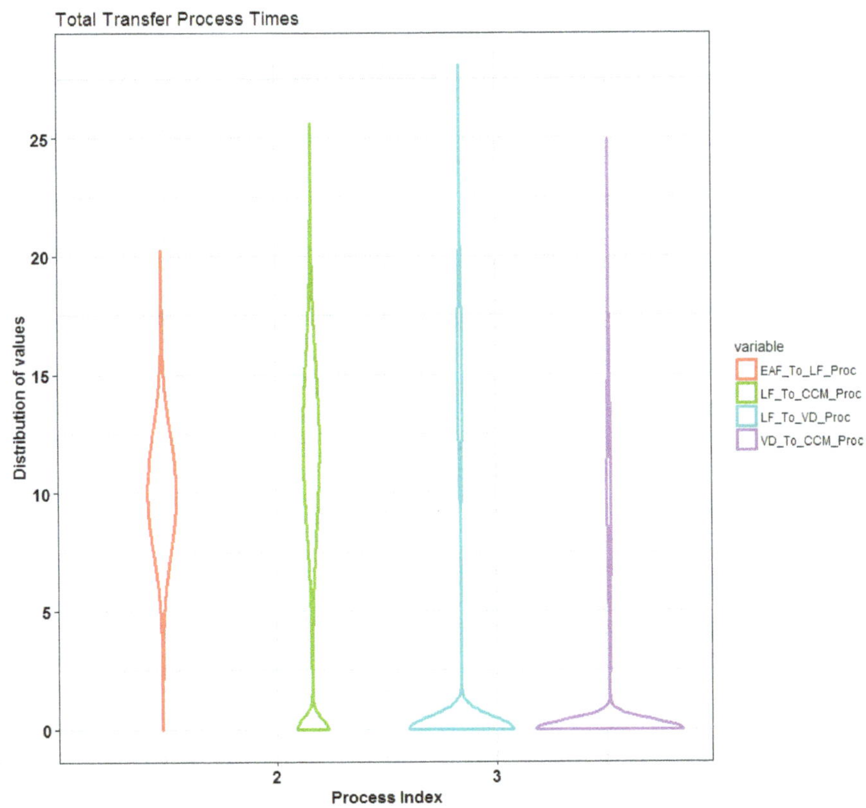

Figure 2. Violin plots of the total transfer-time distributions between processes.

Furthermore, **Figure 3** illustrates the partial process-time distributions of the five metallurgical stations: EAF, LF, VD, SCCM, and BCCM. One may notice that in case that the greatest percentage (80%) of the products is not VD-treated the related process-times distributions are broadly extended. These data sets are also generated during the simulation runs. Based on the ladle refractory maintenance data that are presented in **Table 2**, the simulation program generated the refractory history for the ladle just before EAF tapping in a probabilistic fashion that is illustrated in **Figure 4**. Depending upon the ladle refractory condition a SPH correction as presented in **Table 2** was applied at the LF. Again, here, the great advantage of R upon very fast SPH correction computation should be noted:

$$\begin{aligned} &HeatNr\ <-10000 \\ &SPH\_Corr\ <-replicate(HeatNr,\ get\_Ladle\_SPH\_Correction()) \end{aligned} \tag{4}$$

As described by (4), the vectorization potential of instructions like replicate can perform a computing set of commands—in a function like get_Ladle_SPH_Correction—for a large number of repetitions within a very short period of time. At Stomana meltshop, a great number of grades are produced. In this study, a total of 24 grades for blooms and 6 grades for slabs have been selected. **Figure 5** depicts the average liquidus temperatures based on results that were gathered in the last 17 months. As seen on the graph, the grades are designated in the range of 1–24 for blooms and 51–56 for slabs (**Figure 6**).

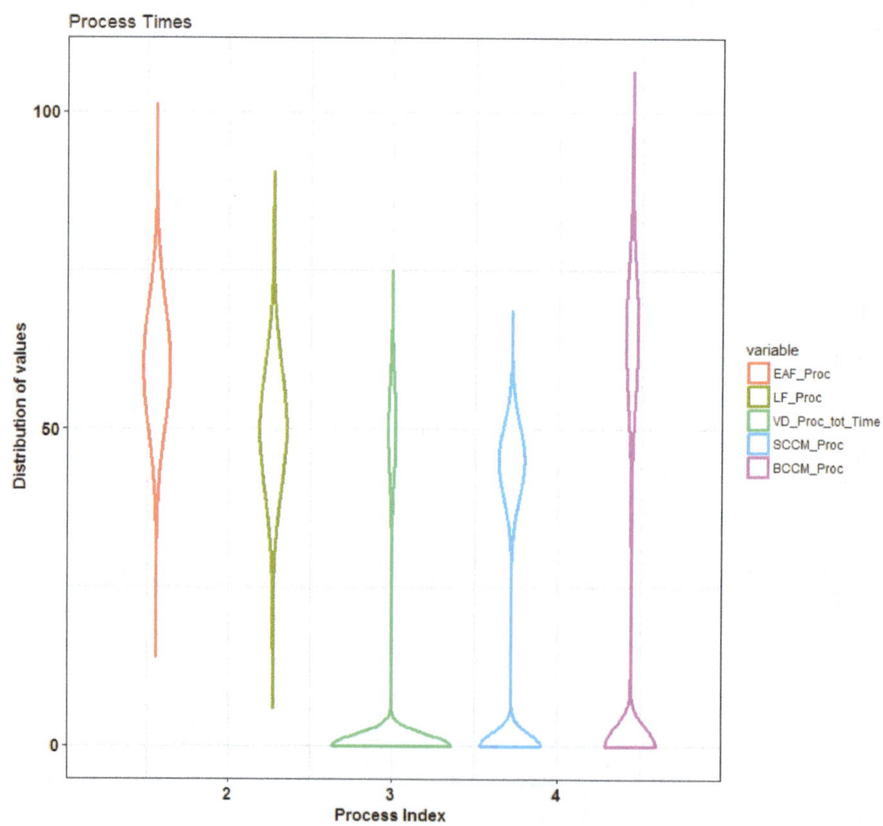

Figure 3. Process-time distributions for the processes: EAF, LF, VD, SCCM, and BCCM.

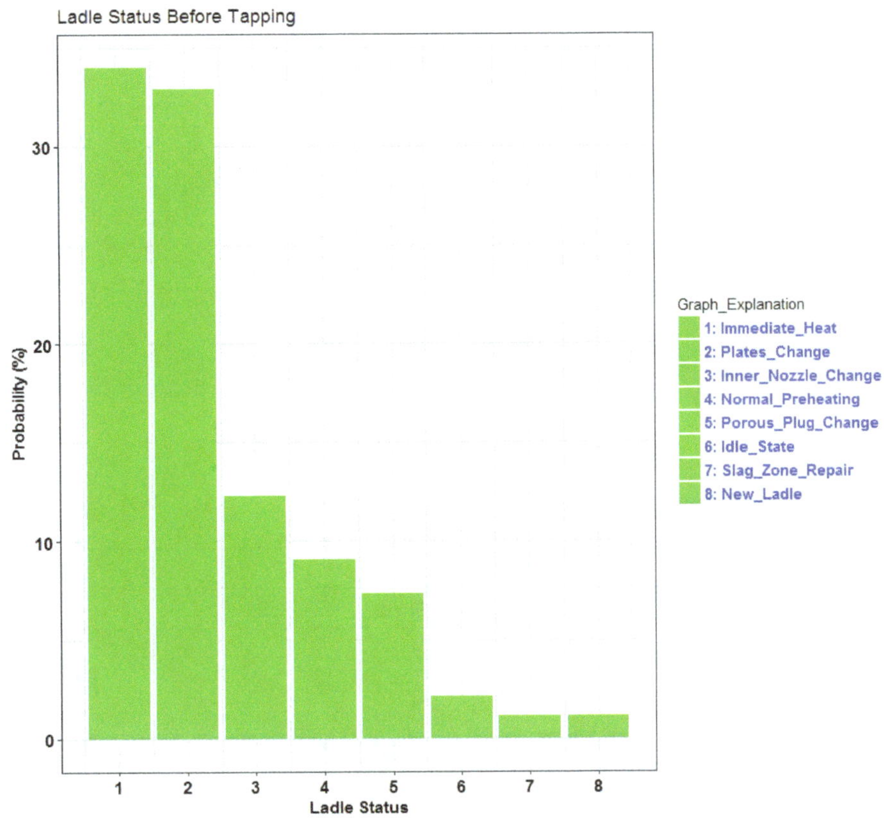

Figure 4. History of ladle refractory just before EAF tapping.

2.2. Deploying the DRF and GBM models

The present chapter was based upon data provided from the Stomana meltshop which is hosted in a steelmaking plant located in Pernik, Bulgaria, and belongs to the SIDENOR/VIOHALCO group of companies; furthermore, another set of data was reproduced by a Monte-Carlo simulation as explained in the previous section. The main task was to generate at least one supervised model that will identify critical parameters that affect the casting floor SPH by adjusting the liquid steel SPH at the LF. The H2O Flow package [19] was deployed for this type of work. This package is available for free from the web, and it is extensively used by many companies and scientific institutions worldwide. Two machine-learning algorithms (models) were used from this package: the distributed random forest (DRF) [20] and the gradient boosting method (GBM) [21]. A GBM is an ensemble of either regression or classification tree models. Both are forward-learning ensemble methods that obtain predictive results using gradually improved estimations. Boosting is a flexible nonlinear regression procedure that helps improve the accuracy of trees. Weak classification algorithms are sequentially applied to the incrementally changed data to create a series of decision trees, producing an ensemble of weak prediction models. While boosting trees increases their accuracy, it also decreases speed and user interpretability. The gradient boosting method generalizes tree boosting to minimize these drawbacks. Finally, the distributed random forest (DRF) is a variation of a general technique called ensemble learning. An ensemble model is composed of the combination of several smaller simple models (often small decision trees). The random forest approach tries to de-correlate the trees by randomizing the set of variables that each tree is allowed to use. The

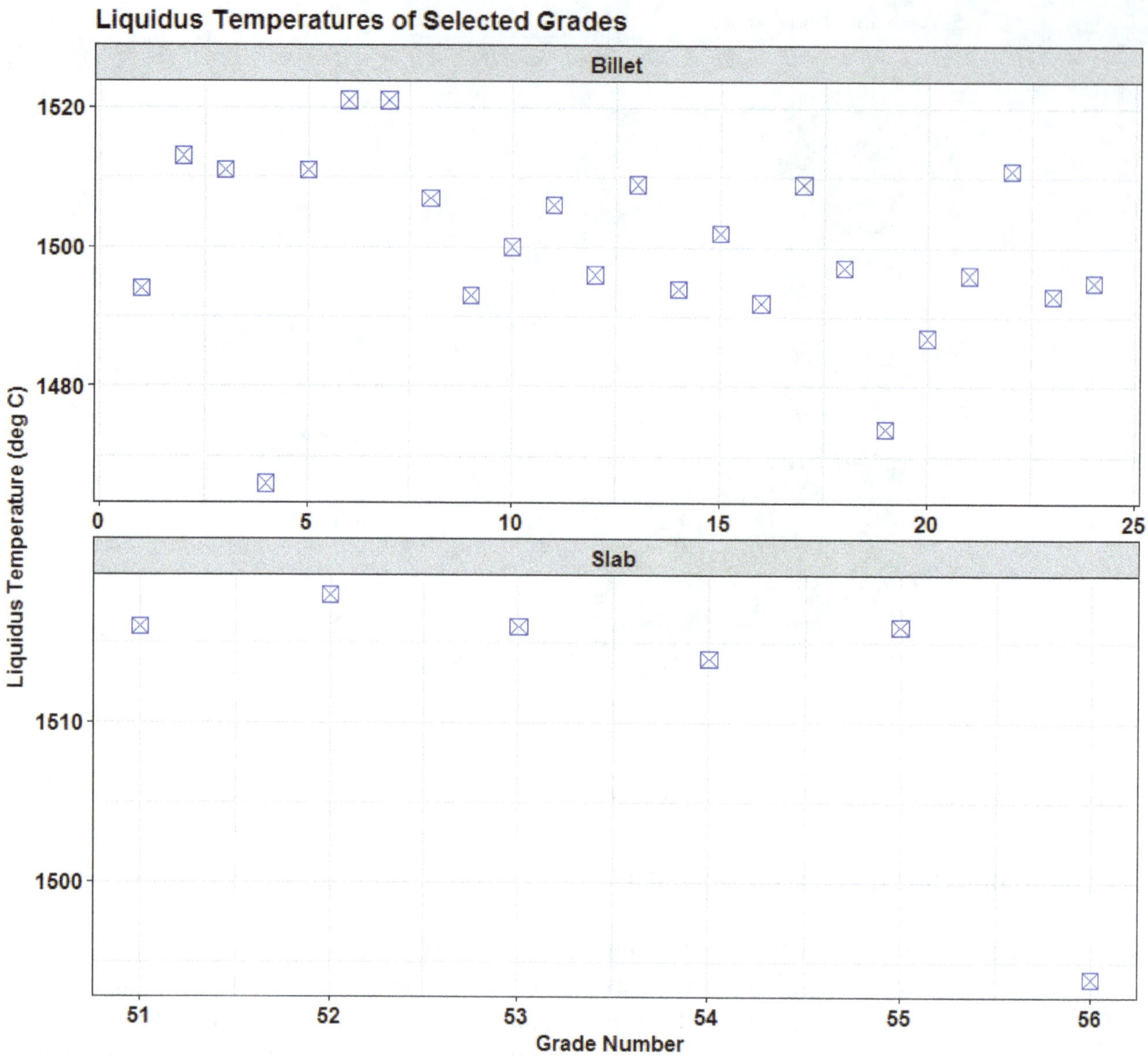

Figure 5. Liquidus temperatures for the selected grades of blooms and slabs.

final ensemble of trees is then bagged to make the random forest predictions [22]. In total, up to 100,000 cases (rows) of data were collected by the simulation software; each case included a heat produced at the EAF, processed at LF, and then directly transferred to the CCM, or after an extra treatment at the VD. The software was run in a DELL Alienware laptop with the Intel i7-6700HQ CPU (8 cores) @2.6 GHz, 16 GB RAM, running under a 64-bit Windows 10 Professional OS. At first, a cluster was generated by Java-Virtual-Machine 64-bit-software called by a program developed for this purpose in R in which the memory size, the number of CPU-cores, and the H2O Flow connection was initialized and established. Then the set of data (data frame) was imported into the cluster. Each time the data frame was split in two frames, in a random fashion: the training data frame consisted of the 75% of the data and the validation data frame consisted of the rest 25%. The models (algorithms) were trained from the 75% of the data and tested (validated) on the rest 25%, generating supervised models that are valid within a measurable statistical error. Two types of running programs were executed per algorithm: in

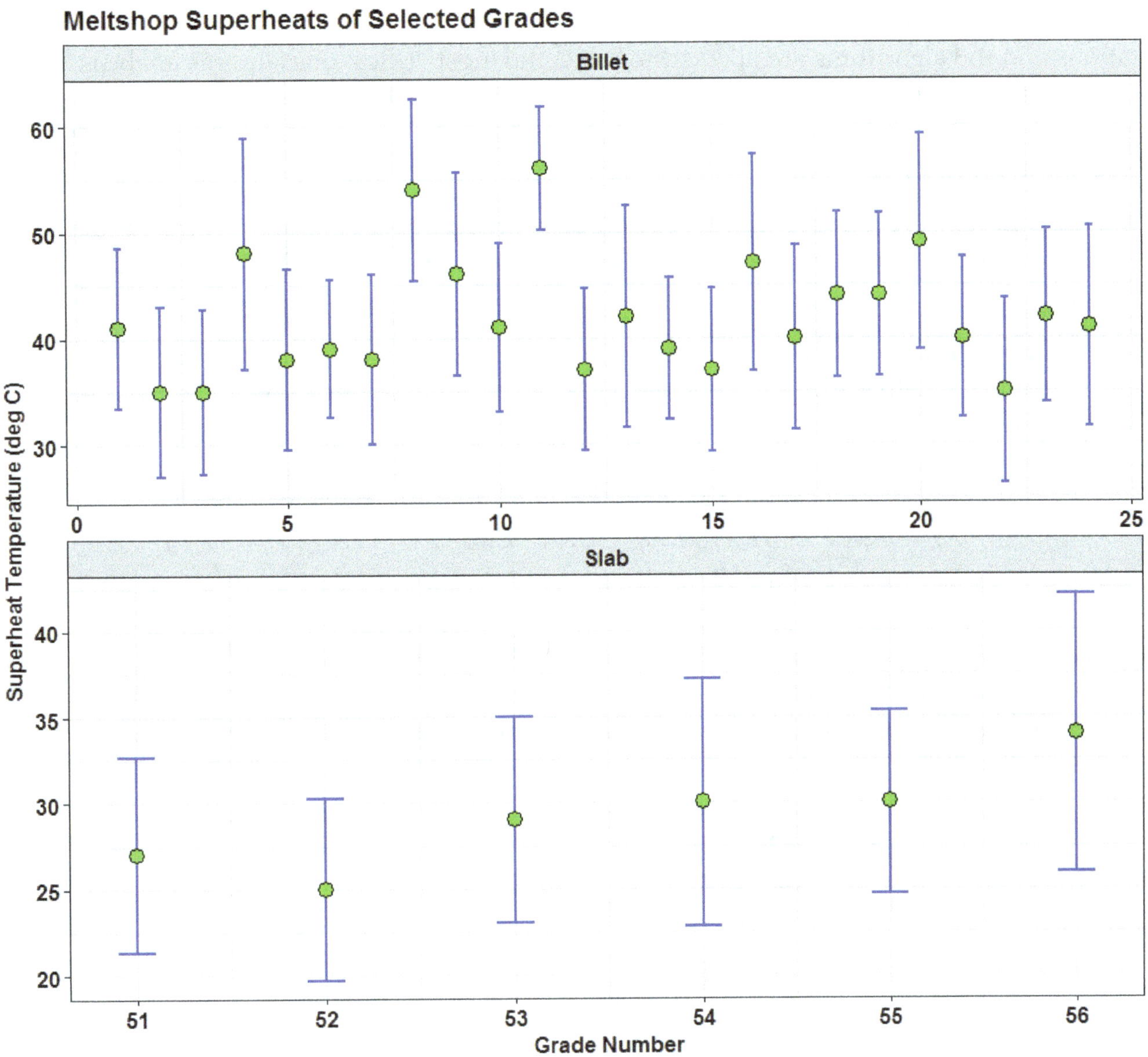

Figure 6. Average SPH values for the selected grades of blooms and slabs based on the current practice; together are presented the limits of plus/minus one standard deviation ($\mu \pm \sigma$).

the first part, a grid search was performed in order to deduce the proper tuning parameters that potentially minimized the validation error, and in the second part, the execution of the tuned model resulted in the derivation of the final supervised model. The grid search is time consuming as it requires a trial-and-error procedure. One final remark concerning the deployment of the H2O Flow package: it may be initiated by R and run in a stand-alone program in R, or run in a web-based framework (e.g., Mozilla Firefox); the latter was extensively used in this study.

3. Results and discussion

Preliminary investigations showed that from the initial set of parameters that were reproduced by the simulation runs, only a few were found critical enough to be included in this type of study. Although, in data-science modeling, all parameters are included in the computations and the algorithms are allowed to select the most critical ones, in this analysis it was considered to decrease the number of most important parameters in order to have the ability to appraise better the phenomena involved. **Table 3** presents the parameters that were finally selected in this part.

The parameter SPH_Overall_ESt was computed based on some assumptions for the temperature loss at the casting floor. **Table 4** presents the values used for the calculation of this term.

The values Cte1, Cte2, etc., used in every simulated test were picked up randomly from a normal distribution with the corresponding (μ, σ) values as shown in **Table 4**; the formula used for parameter SPH_Overall_Est was:

$$\begin{aligned} SPH\_Overall\_ESt \ < - \ SPH\_Corr3 - Cte1 * Holding\_Time- \\ Cte2 * SPH\_HtInSeq\_CORR - Cte3 * VD\_\mathrm{Pr}oc\_tot\_Time- \\ Cte4 * Tund\_Temp\_Drop \end{aligned} \tag{5}$$

The SPH_HtInSeq_CORR term is randomly drafted from a normal distribution of (μ, σ) equal to (15.0, 2.0) for the heats that are cast first in a tundish casting sequence. From practice experience, an extra 15°C temperature is generally required for the first heat in a casting sequence as

Name	Description
SPH_Corr3	The liquid steel SPH at the LF exit
Holding_Time	The time liquid steel is contained in a ladle
VD_Proc_tot_Time	Total processing time of VD process (if any for a heat)
SPH_HtInSeq_CORR	SPH correction if the heat is supposed to be first in a sequence of castings in a tundish
VAR_Grade_Sel	The 30 selected grades for analysis
VAR_Grade_SPH_CCM	The casting floor SPH for the 30 selected grades as experienced in the current actual meltshop practice
SPH_Overall_ESt	The simulated expected/estimated SPH at the casting floor

Table 3. Critical parameters selected for data-science modeling.

Description	Average (μ)	Standard deviation (σ)
Cte1 (Holding_Time, °C/hr)	0.50	0.25
Cte2 (SPH_HtInSeq_CORR)	0.80	0.07
Cte3 (VD_Proc_tot_Time, °C/hr)	1.417	0.133
Cte4 (Tund_Temp_Drop)	0.85	0.05

Table 4. Values used for the calculation of the expected casting floor SPH.

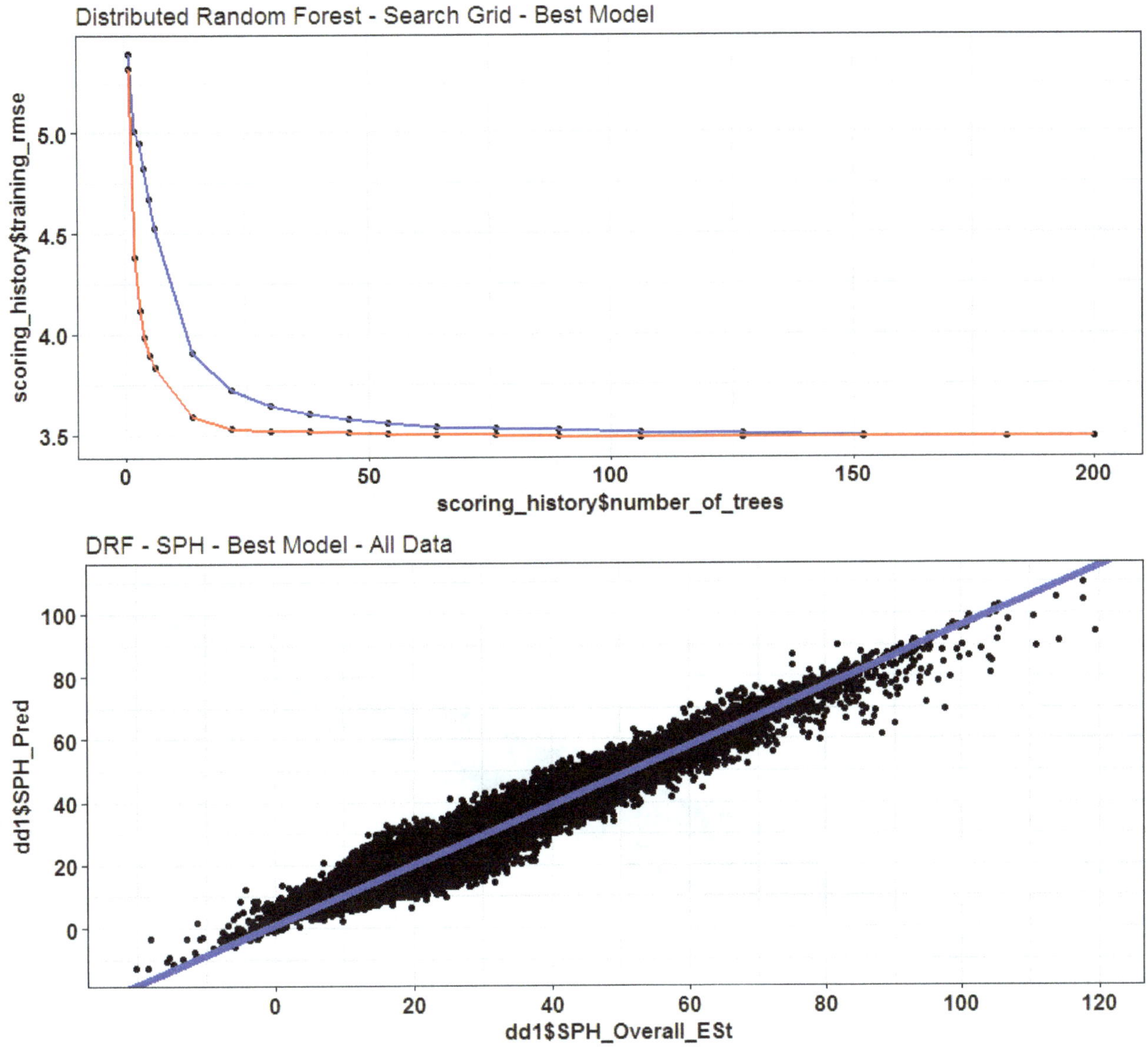

Figure 7. DRF results for the prediction of the casting floor SPH (term SPH_Overall_ESt); top graph presents grid sensitivity analysis in order to select the proper tuning parameters and the bottom graph presents the degree of approximation.

tundish comes from a preheating station at about 1100°C and absorbs some heat from liquid steel. Normally, the ladle-to-tundish liquid-steel transfer operation absorbs some heat; the Tund_Temp_Drop term corresponds to that effect and is also randomly chosen from a normal distribution with (μ, σ) equal to (35.0, 5.0) for all heats. **Figures 7** and **8** illustrate the DRF and GBM results with respect to predicting the SPH_Overall_ESt term.

For both cases, the ANOVA (analysis of variance) [23] gave some good statistical figures; simplifying results for the GBM model only, the residual standard error was 3.159 on 99,998 degrees of freedom, the multiple R-squared was 0.9484, and the F-statistic gave $1.838 \cdot 10^6$ on 1 and 99,998 DF, with a p value $<2.2 \cdot 10^{-16}$. Normally, the GBM algorithm suffices to come up with a reasonable supervised model; however, the DRF algorithm was added for comparison purposes.

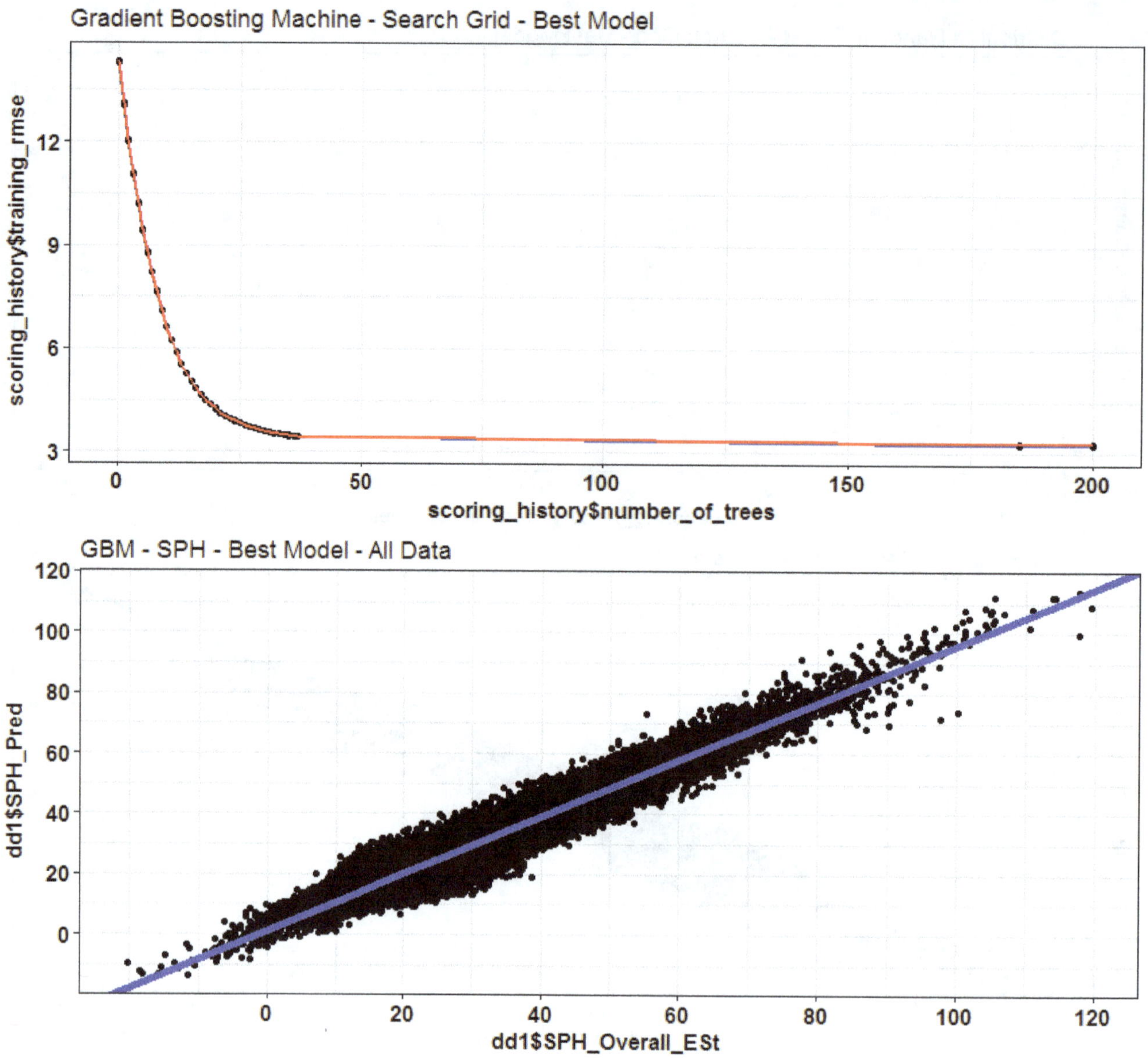

Figure 8. GBM results for the prediction of the casting floor SPH (term SPH_Overall_ESt); top graph presents grid sensitivity analysis in order to select the proper tuning parameters, and the bottom graph presents the degree of approximation.

Order	Parameter
1	SPH_Corr3
2	VD_Proc_tot_Time
3	SPH_HtInSeq_CORR
4	Holding_Time
5	VAR_Grade_SPH_CCM
6	VAR_Grade_Sel

Table 5. Relative importance of variables for the prediction of SPH_Overall_ESt (GBM model).

Table 5 shows the relative importance of the considered parameters for the prediction of SPH_Overall_ESt given by the GBM model; the recommended LF-exit SPH (SPH_Corr3) plays a great role, indeed. Ignoring the SPH_Overall_ESt term, one interesting analysis could be the prediction of the current practice superheats (actual SPH, term VAR_Grade_SPH_CCM) at the casting floor for the selected grades; it should be pointed out that the selection of these grades is completely at random, that is, the simulated heats do not follow at all the SPH data from the current meltshop practice. Nevertheless, the deduced DRF and GBM supervised models exhibited a remarkable statistical significance: again, simplifying results for the GBM model only, the residual standard error was 4.988 on 99,998 degrees of freedom, the multiple R-squared was 0.5352, and the F-statistic was $1.152{\cdot}10^5$ on

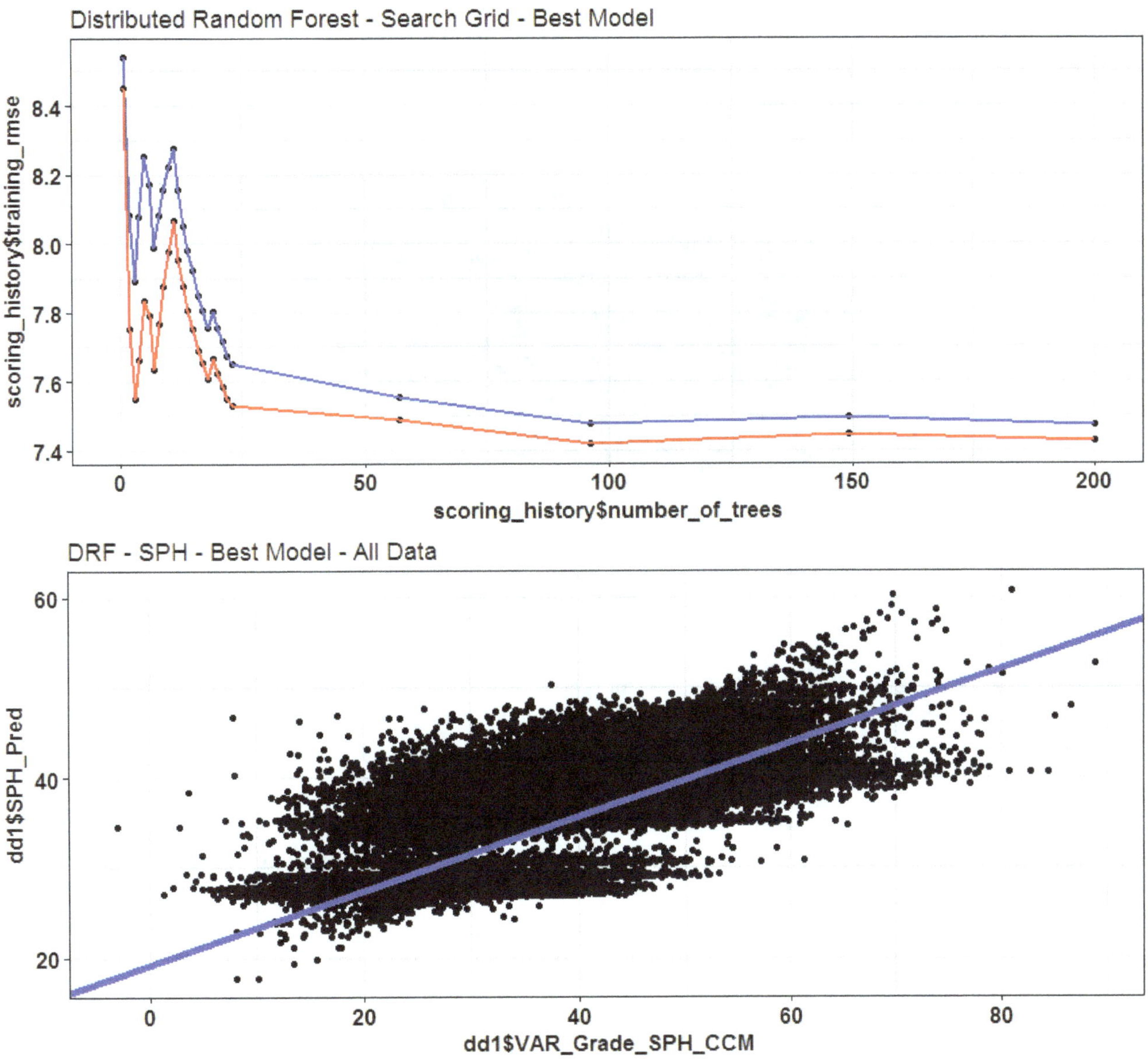

Figure 9. DRF results for the prediction of the current practice casting floor SPH (term VAR_Grade_SPH_CCM); top graph presents grid sensitivity analysis in order to select the proper tuning parameters, and the bottom graph presents the degree of approximation.

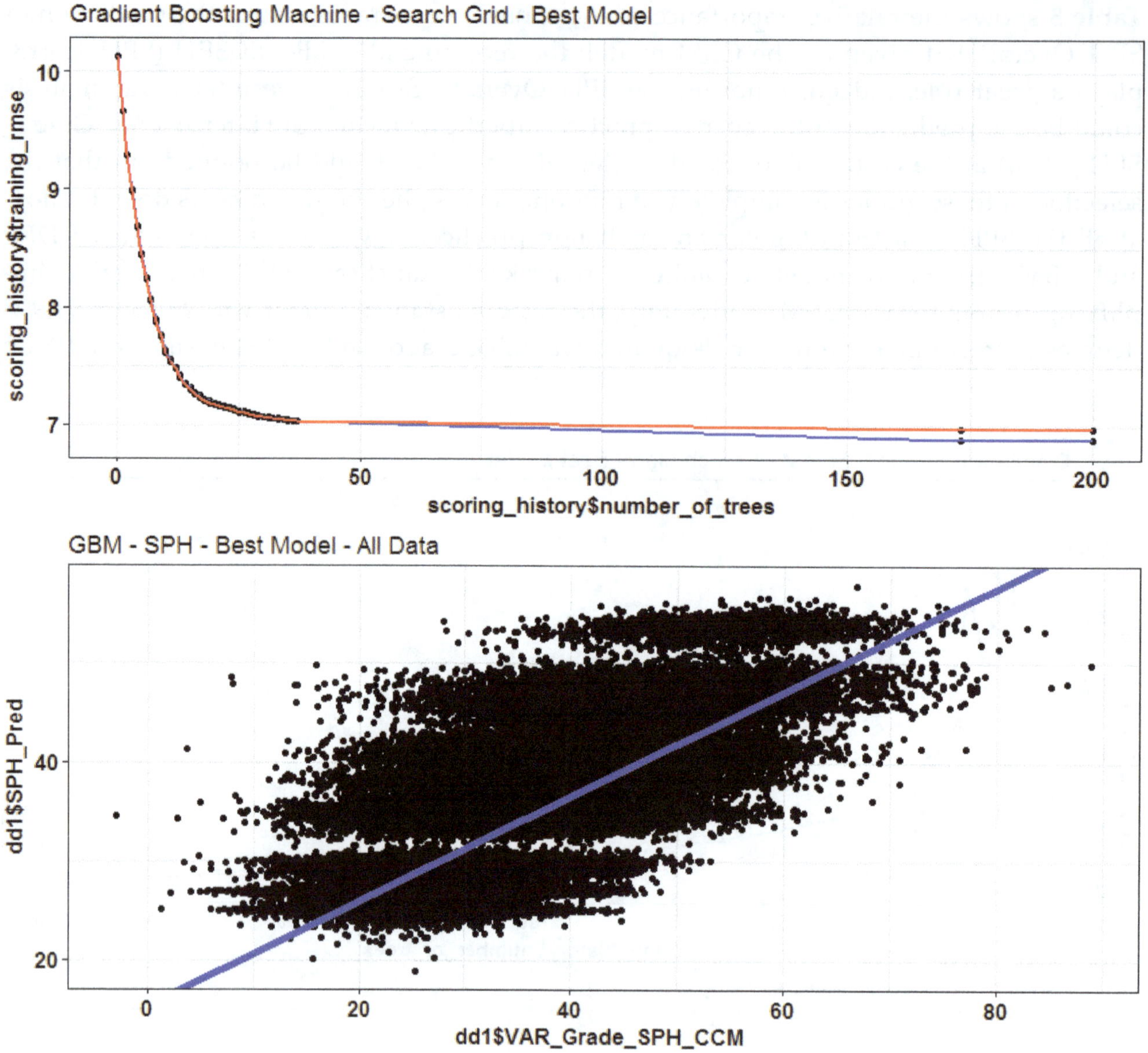

Figure 10. GBM results for the prediction of the current practice casting floor SPH (term VAR_Grade_SPH_CCM); top graph presents grid sensitivity analysis in order to select the proper tuning parameters, and the bottom graph presents the degree of approximation.

1 and 99,998 DF, with a p-value $<2.2 \cdot 10^{-16}$. **Figures 9** and **10** illustrate these findings for the prediction of the important term VAR_Grade_SPH_CCM by excluding the computed term SPH_Overall_ESt.

Table 6 illustrates the relative importance of the parameters that were considered for the prediction of the current practice superheats (VAR_Grade_SPH_CCM) given by the GBM model. The great importance of the selected grade parameter (VAR_Grade_Sel) seems as expected due to the nature of this supervised model; however, SPH_Corr3 still appears to be very important. Apart from the analysis so far, one extra step was taken in order to test whether the derived results may be attributed to pure coincidence. In the position of the SPH_Overall_ESt

Order	Parameter
1	VAR_Grade_Sel
2	VD_Proc_tot_Time
3	SPH_Corr3
4	Holding_Time
5	SPH_HtInSeq_CORR

Table 6. Relative importance of variables for the prediction of VAR_Grade_SPH_CCM (GBM model).

term, the term SPH_tun1 was placed. This resembles more to the initial tundish temperature (1) of the Gupta et al [7] work, that exhibited a correlation coefficient $R^2 = 0.73$; indeed, after some manipulation the following equation was derived:

$$SPH\_tun1 = f + 0.019LL - 0.012TAT - 0.011T_{liq} + 0.358SPH_{past} + 0.631SPH_{LF} \quad (6)$$

One should recall that for the term SPH_{LF} the known term SPH_Corr3 can be used. T_{liq} is the liquidus temperature of the selected grades, and f is a function of the Holding_Time. The T_{liq} and SPH_{past} terms were randomly gathered from normal distributions with (μ, σ) equal to (1490.0, 10.0) and (40.0, 5.0), respectively. **Figure 11** illustrates the derived GBM supervised model for the prediction of the SPH_tun1 term as computed in (6).

The ANOVA for the model results presented in **Figure 11** exhibited the following statistical significance: the residual standard error was 2.446 on 56,189 degrees of freedom, the multiple R-squared was 0.9659, and the F-statistic was $1.589 \cdot 10^6$ on 1 and 56,189 DF, with a p-value $<2.2 \cdot 10^{-16}$. In **Table 7**, the recommended LF-exit superheat (SPH_Corr3) still appears to be of great importance.

Although 100,000 heats were simulated, a number of data had to be excluded from the data-science analysis in case that some SPH_tun1 predictions were outside the (10.0, 70.0) range. The statistical significance appears to be more than satisfactory, realizing that the parameters presented in **Table 3** were taken under consideration with the only substitution of term SPH_tun1 in the place of term SPH_Overall_ESt. One final thing has to be mentioned: normally, Monte-Carlo type simulations converge to an average value (μ) and a standard deviation (σ) that tends to decrease as the number of repetitions (number of heats in this case) increases. **Figure 12** describes these findings by simulating meltshop production from 1000 till 250,000 heats. The computed SPH values for $\mu + 3^*\sigma$ exhibit a tendency to decrease as the number of heats increases. At the same time, the reduction of the expected SPH values, as the number of heats increases, seems to point out that there is a tendency for improvement once some logic is involved in the recommendation of LF exit SPH temperatures.

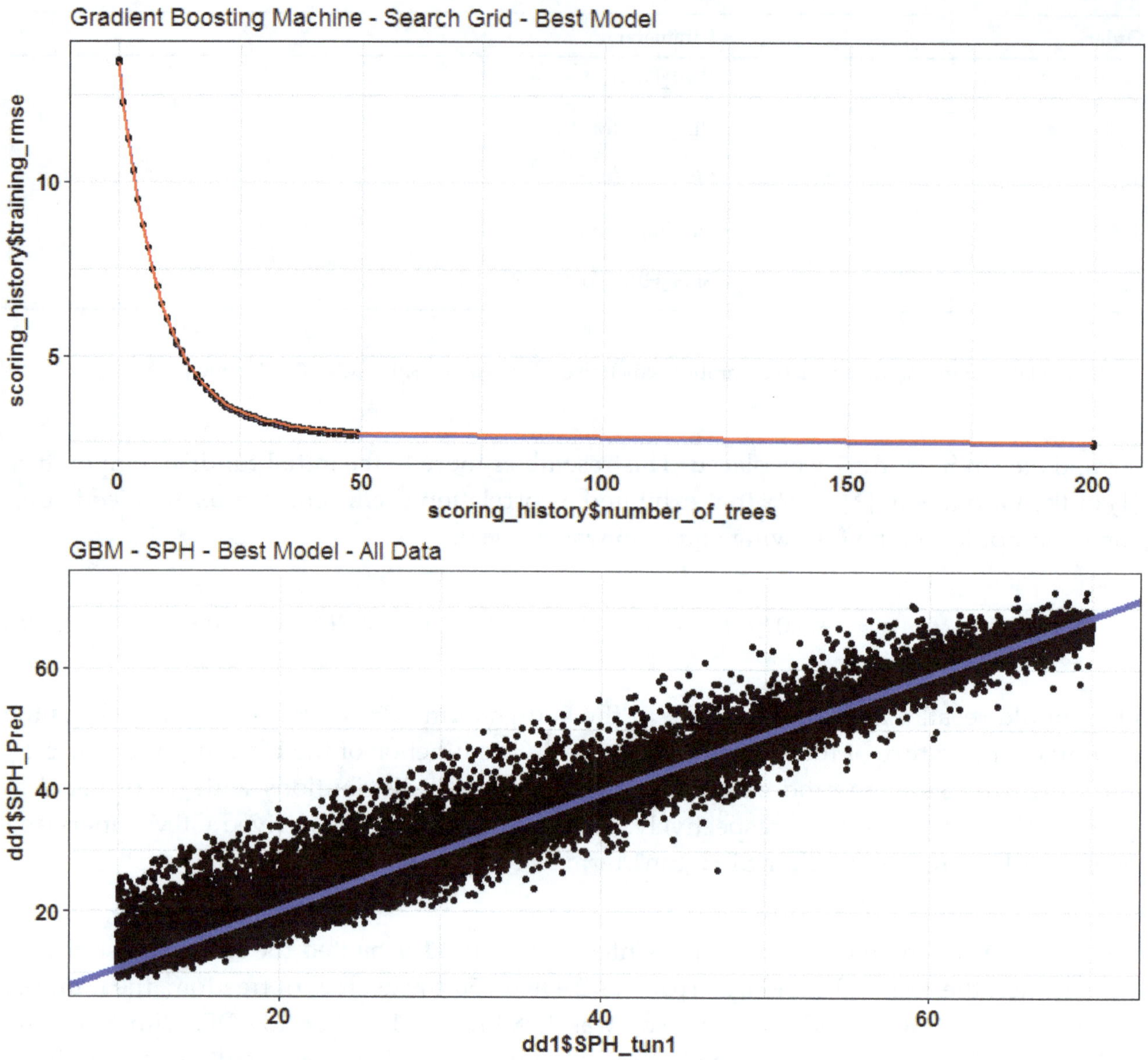

Figure 11. GBM results for the prediction of the first SPH at the casting floor based on Gupta et al. [7] (term SPH_tun1); top graph presents grid sensitivity analysis in order to select the proper tuning parameters, and the bottom graph presents the degree of approximation.

Order	Parameter
1	Holding_Time
2	SPH_Corr3
3	VD_Proc_tot_Time
4	SPH_HtInSeq_CORR
5	VAR_Grade_SPH_CCM
6	VAR_Grade_Sel

Table 7. Relative importance of variables for the prediction of SPH_tun1 (GBM model).

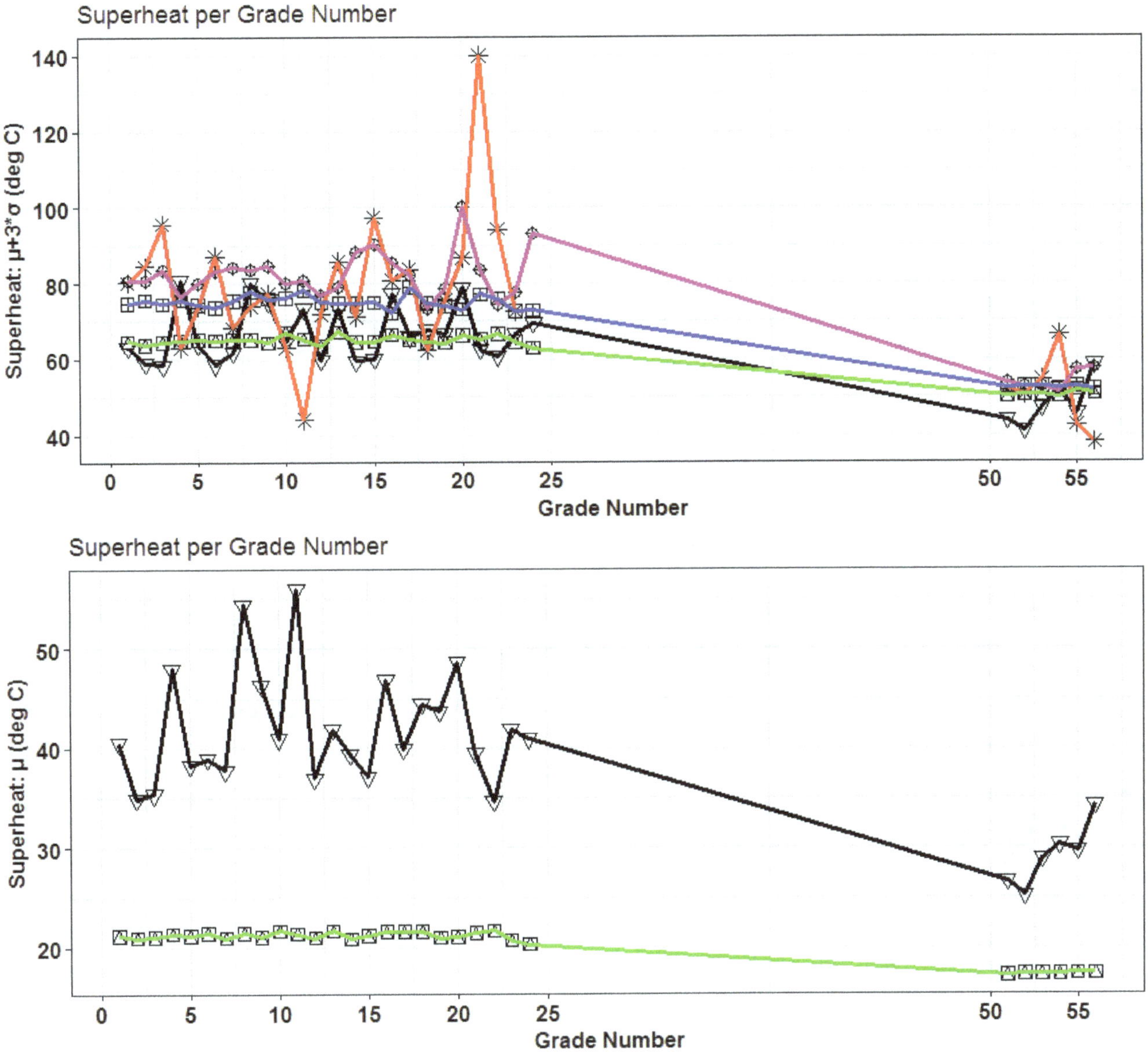

Figure 12. Potential of improvement on the SPH per grade. Top graph: maximum values (μ + 3*σ) of SPH (term SPH_Overall_ESt) with respect to simulated number of heats; 1000 heats (1), 10000 heats (2), 100000 heats (3, dotted), 250000 heats (4, dashed), current SPH practice (5, solid). Bottom graph: current average SPH values (solid), expected SPH (SPH_Overall_ESt) values (dashed).

4. Conclusions

A Monte-Carlo simulation software was developed in order to reproduce meltshop data concerning process times, ladle refractory history, and effect on liquid-steel temperature loss at the casting floor. Data-science modeling was applied in order to deduce supervised algorithms for the prediction of casting floor superheats based on critical parameters from reproduced and plant data. The results were also related with findings from a published work. In most cases, the derived supervised models exhibited a remarkable statistical significance, which seems to be too difficult to occur due to pure coincidence. It is very likely that a ladle tracking system will greatly result in a better achievement of desired casting floor superheats, and therefore, important economic savings.

Acknowledgements

The author is grateful to the top-management support in this study.

Author details

Panagiotis Sismanis

Address all correspondence to: psismanis@sidenor.vionet.gr

SIDENOR SA, Athens, Greece

References

[1] Ayata K, Mori H, Taniguchi K, Matsuda H. Low superheat teeming with electromagnetic stirring. Iron and Steel Institute of Japan International. 1995;**35**(6):680-685

[2] Thomas B. Issues in thermal-mechanical modeling of casting processes. Iron and Steel Institute of Japan International. 1995;**35**(6):737-743

[3] Guyot V, Martin JF, Ruelle A, d' Anselme A, Radot JP, Bobadilla M, Lamant JY, Pontoire JN. Control of surface quality of 0.08% < C < 0.12% steel slabs in continuous casting. Iron and Steel Institute of Japan International. 1996;**36**:S227-S230

[4] Jansto SG. Steelmaking and continuous casting process metallurgy factors influencing hot ductility behavior of niobium bearing steels. Metal. Brno, Czech Republic; 2013. pp. 36-42

[5] Jacobi H, Schwerdtfeger K. Ripple marks on cast steel surfaces. Iron and Steel Institute of Japan International. 2013;**53**(7):1180-1186

[6] Smithells CJ. Metals Reference Book. 5th ed. London: Butterworths; 1976

[7] Gupta N, Chandra S. Temperature prediction model for controlling casting superheat temperature. Iron and Steel Institute of Japan International. 2004;**44**(9):1517-1526

[8] Addes VI, Sabol JD. Development and implementation of the process model for controlling casting superheat temperature. In: Steelmaking Conference Proceedings. Iron & Steel Society; 1996. pp. 333-340

[9] Fredman TP, Torrkulla J, Saxen H. Two-dimensional dynamic simulation of the thermal state of ladles. Metallurgical and Materials Transactions B. 1999;**30B**:323-330

[10] Tian H, Mao Z, Wang Y. Hybrid model of molten steel temperature prediction in LF. Iron and Steel Institute of Japan International. 2008;**48**(1):58-62

[11] Tian H, Mao Z, Wang A. New incremental learning modeling method based on multiple

models for temperature prediction of molten steel in LF. Iron and Steel Institute of Japan International. 2009;**49**(1):58-63

[12] Chen S, Abraham S. On-line superheat control model for continuously cast slabs and billets. Iron & Steel Technology. 2010;**7**:89-96

[13] Chen S, D'Souza C, Evans D, Dunnett K, Burns J, Sylvestre G, Cannon C. Continuous enhancement of the EVRAZ superheat model control for slab casting. Iron & Steel Technology. 2013;**7**:85-96

[14] Sonoda S, Murata N, Hino H, Kitada H, Kano M. A statistical model for predicting the liquid steel temperature in ladle and Tundish by bootstrap filter. Iron and Steel Institute of Japan International. 2012;**52**(6):1086-1091

[15] AustralTek. Physical ladle tracking system [Internet]. Available from: http://steeltracking.com/video.html [Accessed: November 5, 2017]

[16] SinterCast CGI. SinterCast tracking technologies [Internet]. Available from: http://sintercast.com/technology/sintercast-tracking-technologies [Accessed: October 20, 2017]

[17] Hillier FS, Lieberman GJ. Operations Research. 2nd ed. San Francisco: Holden-Day Inc; 1974

[18] R Foundation. The R project for statistical computing [Internet]. Available from: https://www.r-project.org/ [Accessed: August 30, 2017]

[19] H2O.ai. H2O flow [Internet]. Available from: https://www.h2o.ai/h2o/h2o-flow/ [Accessed: August 31, 2017]

[20] Aiello S, Eckstrand E, Fu A, Landry M, Aboyoun P, Lanford J, editors. Machine learning with R and H2O [Internet]. [Updated: January 2017]. Available from: http://h2o-release.s3.amazonaws.com/h2o/rel-tutte/2/docs-website/h2o-docs/booklets/RBooklet.pdf [Accessed: November 5, 2017]

[21] Click C, Malohlava M, Candel A, Roark H, Parmar V, Lanford J, editors. Gradient boosted models with H2O [Internet]. [Updated: February 2016]. Available from: https://h2o-release.s3.amazonaws.com/h2o/rel-tukey/6/docs-website/h2o-docs/booklets/GBM_Vignette.pdf

[22] Zumel N, Mount J. Practical Data Science with R. New York: MANNING; 2014

[23] Faraway JJ. Practical Regression and ANOVA with R. Bath: University of Bath; 2002

8

Continuous Anything for Distributed Research Projects

Simon Volpert, Frank Griesinger and
Jörg Domaschka

Abstract

International research projects involve large, distributed teams made up from multiple institutions. These teams create research artefacts that need to work together in order to demonstrate and ship the project results. Yet, in these settings the project itself is almost never in the core interest of the partners in the consortium. This leads to a weak integration incentive and consequently to last minute efforts. This in turn results in Big Bang integration that imposes huge stress on the consortium and produces only non-sustainable results. In contrast, industry has been profiting from the introduction of agile development methods backed by Continuous Delivery, Continuous Integration, and Continuous Deployment. In this chapter, we identify shortcomings of this approach for research projects. We show how to overcome those in order to adopt all three methodologies regarding that scope. We also present a conceptual, as well as a tooling framework to realise the approach as Continuous Anything. As a result, integration becomes a core element of the project plan. It distributes and shares responsibility of integration work among all partners, while at the same time clearly holding individuals responsible for dedicated software components. Through a high degree of automation, it keeps the overall integration work low, but still provides immediate feedback on the quality of the software. Overall, we found this concept useful and beneficial in several EU-funded research projects, where it significantly lowered integration effort and improved quality of the software components while also enhancing collaboration as a whole.

Keywords: Continuous Delivery, Continuous Integration, Continuous Deployment, project management, software quality, DevOps, distributed software development

1. Introduction

The rise of agile software engineering strategies has leveraged the realisation of Continuous Integration proposed by Grady Booch as early as 1991 [18]. Continuous Integration propagates a constant integration of changes to code as opposed to Big Bang integration done at the end of a development cycle. It has, in turn, paved the path to Continuous Delivery and Continuous Deployment of software components and entire software platforms. The core idea of Continuous Delivery is to be able to roll out new releases at any time and not only at the end of larger development cycles. In order to reach this goal, Continuous Delivery demands the automation of all steps required to compile, bundle, test, and release the software. Testing ranges from unit tests targeting a single software component, over integration tests, acceptance test to user acceptance tests. While this approach is emergently successful in industry, it is barely used for scientific software, neither is it used in collaborative research environments.

In contrast to industrial projects and even open source software projects, distributed research projects (for instance, large(r) EU-funded ICT projects), the project itself is almost never in the core interest of the partners forming the project consortium. Instead, every partner is interested in the niche aspect that made him join the project and that makes the consortium look complete and gives the consortium a complementary appearance. In reality, though, the agendas of the project partners are often driven by their individual interests and particularly academic partners do have an obvious interest in the actual research aspect of the work and are less focussed on the provisioning of dependable, sustainable software artefacts. Neither are they preliminary interested in the common, integrated, stable software platform. In practise, this may mean that Partner A wants to improve on an algorithm they have, while Partner B would define a domain specific language (DSL) for a specific scope and Partner C will provide an improved kernel module for handling I/O on solid state disks. From a research point of view, this means that the main work for Partner A will be on the definition of the algorithm, its implementation, and evaluation in some limited, publishable scope. For Partner B, the main work will be on the definition of the DSL and on applying and realising it in two or three use case scenarios. Partner C's work will be on the definition of the new approach and on the realisation and evaluation of the kernel module, probably for one specific version of Linux.

While the behaviour of all three partners is fully legitimate and understandable, it is the nature of a distributed research project that interdependencies between parts of the software exist. Usually software integration is required at certain project milestones where prototypes should be released and new, emergent features be demonstrated to the public or at least the funders. Here, the lack of common interest in the project in combination with the described "research style" code quality makes the integration a painful, cumbersome, and frustrating task. Our experience shows that in many projects the task of integrating software from different partners is outsourced in an own project work package and then assigned to one or at most two partners that were not or only marginally involved in improving the algorithm from Partner A, developing the DSL from Partner B, and realising the kernel module of Partner C. Furthermore, in many projects the whole integration of all software components is done in a Big Bang style before a review or before an obligatory software release and even worse often performed by a single individual. This poor devil ends up integrating and fixing several dozen software components (s)he has not developed, is not owning, and has never been responsible for.

We argue that instead of putting all integration responsibility and work on the shoulders of a single individual, it is way better to spread out work among project partners and make it everybody's task. We further believe that the techniques and strategies offered by **Continuous Integration**, **Continuous Delivery**, and **Continuous Deployment** are beneficial for enforcing the distribution of the task, automating the necessary steps, monitoring the status of, and gaining confidence in the produced software. The main contribution of this chapter is a concise technical and organisational framework for Continuous Integration, Continuous Delivery, and Continuous Deployment in distributed (research) projects. The framework is based on our experience in half a dozen mid-sized EC projects of 5–25 partners and several smaller sized national-funded research projects. It fairly distributes work among partners and improves overall code quality.

The rest of this chapter is structured as follows. Section 2 identifies the requirements in more detail and presents related work. Sections 3–5 introduce background on Continuous Integration, Continuous Delivery, and Continuous Deployment, respectively. Section 6 presents our framework on both conceptual and tooling level while Section 7 discusses the approach. Section 8 concludes and gives an outlook on future work.

2. Problem statement and related work

International ICT research projects involve large, distributed teams (consortia) made up from multiple companies or institutions, so called partners or beneficiaries. These teams create research software artefacts that need to work together in order to demonstrate and ship the project results. In the following, we analyse the challenges of such constellations, and why this requires a special integration strategy. Finally, we carve out the requirements towards such an integration strategy and discuss related work.

2.1. Challenges with distributed research teams

Development in distributed, that is, non co-located, teams is challenging, as the distribution aspect hinders communication. For instance, meetings and synchronisation actions barely can happen in a timely or even ad hoc manner, causing delays. From a technical point of view this may lead to diverging developments at different locations. From an organisational point of view, it causes overhead.

Koetter et al. [10] identify major problems in distributed teams and particularly with respect to research projects. At the core of their analysis, they identify the team distribution and lacking stakeholder commitment as major problems. The former complicates team communication leading to a lack of internal communication. The latter, a consequence of diverging goals and different (research) interests, leads to a lack of incentives for prototype integration.[1] The impact of these causes is further increased by different cultural and technical background, etiquette, company policies, and high personal fluctuation in research projects.

In order to cope with the diversity and resulting centrifugal forces, it is common that project management applies rules: these range from a common toolset and document template to regular virtual and physical meetings; both intended to improve communication. Additionally,

[1]Please note that "integration" as in "Continuous Integration" has a different meaning than "prototype integration". As defined later, the first tackles integration on code level, while the later addresses the integration of distributed software architecture. In order to avoid misunderstandings, we will always use the full terms.

most projects announce a central technical responsible whose role is to break ties in technical discussions. Finally, technical work is often separated such that local teams at partner sites work independently on certain sub topics producing isolated assets.

From our experience, these measures usually work fine and minimise the tension in the consortium. The lack of common goals usually gets masked by introducing a storyline every partner can agree on. Yet, we claim that the only aspect that cannot be handled by these measures is the work to be done for prototype integration, because it requires that components developed in isolation, work together smoothly despite the weak communication, and common goals. The often-practised Big Bang integration of artefacts causes a lot of work, troubles the consortium, and results in poor software quality.

2.2. The cause for Continuous Anything

Understanding and accepting that Big Bang integration causes pain and sub-optimal results, leads to the insight that a different prototype integration strategy is needed. Ironically, software development industry was facing similar issues decades ago [16] which led to abandoning of the waterfall model and the introduction of so called agile development methodologies. These were stated in the agile manifesto [17] and are being realised by methodologies such as extreme programming, Scrum, or Kanban.

All of these methodologies assume co-located teams with large common interests and a high intrinsic motivation to deliver high quality, usable software. In consequence, they cannot be applied directly to research projects that do not fulfil the necessary preconditions. Nonetheless, at the core of their prototype integration[2] methodology, agile methodologies rely on a highly automated, frequently executed, and constant process to reduce the possibility for human errors and to obtain continuously executable software artefacts.

While such an approach requires an upfront and constant invest in prototype integration, the overall amount of effort needed per partner and particularly per consortium is likely to be a lot less compared to Big Bang integration. This is due to the fact that changes are small and can be easily reviewed. Moreover, the use of automation allows dealing with the complexity of even larger and more diverse teams.

2.3. Constraints and requirements

We claim that automation can reduce the pain for prototype integration in (large) research projects. Yet, as with improving communication within the consortium, introducing an automated process, this improvement will not happen for free. Work from the project management is needed to establish and enforce such a process, which may cause resistance.

Therefore, our major aim is to minimise the upfront investment of project partners and the management effort needed to enforce the strategy. The overall goal is to develop an automated prototype integration schema that takes into account the specific needs of research consortia. This is broken down into particular requirements and challenges presented in the following sections.

[2]For industry it should rather be "product integration".

2.3.1. General requirements

This section covers requirements towards automating prototype integration. We present them from the perspective of a research project, but they can be applied in different settings.

R.1 (distribute work among partners): As partners do not have common interest and no incentives for prototype integration, it is necessary to not burden one team or even one individual with this task, but to achieve a fair distribution of work among partners.

R.2 (reduce manual burden): Integration work distracts researchers from their work and so does testing. Due to that, as much as possible of the prototype integration work and testing should be automated. This includes reporting if code is currently working or not.

R.3 (denote responsible persons): With high automation degree and the capability to identify non-working portions of the code, denoting responsible persons for individual software components, libraries, and features, allows explicitly tasking those for fixing the non-working parts of the architecture.

R.4 (make the product easy to start and use): Having a project outcome that is easy to install, to start, and to demonstrate, tremendously reduces the burden when planning for a review, a demo, or a webinar.

R.5 (clarify big picture and software dependencies): In large software projects, it is often the case that the big picture is forgotten. In particular, in research projects, researchers lose themselves in details of research questions. Hence, it is important to keep an eye on the overall architecture and ensure that the interactions between components work as intended.

R.6 (make the software status visible): Making the product (including its sub-products) visible helps consortium members to understand what the others are doing. It also helps use case partners giving feedback on the project and the work currently done.

2.3.2. Specific challenges of research projects

Besides the generic requirements that can be found in many distributed teams, the fact that distributed research projects are often executed by loosely coupled beneficiaries creates further technical challenges.

R.A (cater for closed code or even unavailable source code and binaries): Due to different commercial interests of partners, it may be the case that some of them release source code only in a restricted manner or not at all. In some cases, partners do not even release binaries to the rest of the consortium. The overall prototype integration strategy has to be able to deal with this.

R.B (support fluctuation of team members): Research projects have a very high fluctuation of team members. As the budget is fixed, it happens that more people come into the project towards the end, if budget is still available. On the other hand, if the project is short on budget, expensive, that is, senior researchers are moved away and juniors or even undergrads join. This forbids that there are hidden, that is, implicit, agreements between individuals.

R.C (keep track of targeted outcomes): The fluctuation of team members and the dynamic of IT research lead to often changing technical goals of the research project. As a consequence, the current goals need to be documented and be accessible by all project members in order to keep a common focus. Ideally, they are immediately visible to all contributors.

R.D (support different configurations): Often research projects do not build generic solutions, but only demonstrators for specific use cases. The prototype integration strategy has to be able to deal with this and provide the capability to set-up different environments.

R.E (allow different programming languages and development methodologies): Research projects barely start from scratch, as many partners continue earlier work. Further, the knowledge and suitability of languages is very specific to problem domains. Therefore, a prototype integration strategy has to be open to different programming languages.

2.4. Approach

In industrial contexts solving integration and communication issues is realised by introducing three kinds of orthogonal, but complementary approaches: *Continuous Integrati*on handles the integration on code level per component. It builds and tests the component whenever a new version of the code is available. *Continuous Delivery* is concerned with taking the new version of a component and packing it in a shippable box. In addition, it runs integration tests to ensure the interplay with other components. Finally, *Continuous Delivery* takes the component and installs it in a pre-defined environment.

In Sections 3–5, we show that an adapted process to Continuous Integration, Continuous Delivery, and Continuous Deployment can indeed overcome integration issues for distributed research projects. In addition, Section 6 presents a set-up that is able to deal with the requirements from Section 2.3.

2.5. Related work

While there is a lot of literature on DevOps [15], agile methods, Continuous Integration, Continuous Delivery [13], and Continuous Deployment not much can be found with respect to academia and academia/industry collaboration. Eckstein provides guidelines for distributed teams [11].

Rother [1] lays part of the foundation of what is today perceived as DevOps by presenting the production pipelines and methodologies of Toyota. Being more on the cultural side of DevOps and CI/CD spectrum, Davis and Daniels [14] and Sharma [12] give some insights on how to bring these ideas to industry.

Especially Continuous Integration was significantly influenced by Duvall et al. [2]. There, the authors describe most of the paradigms important for Continuous Integration. These are still valid today and are considered as de-facto standard. Fowler's influential articles on Continuous Integration, for example [5], and testing, for example, through micro-service scenarios [4], lay the foundation on what is being perceived as Continuous Integration along with best practices.

Regarding academia, there is ongoing effort in bringing Continuous Integration and Continuous Delivery to teaching. Eddy et al. [7] describe how they implement a pipeline for supporting their lecture on modern development practices. An academic view on Continuous Experimentation is brought up by Fagerholm et al. [9] by investigating multiple use-cases of industry partners. They analyse the demands and propose solutions to create experimentation-happy environments utilising Continuous Integration and Continuous Delivery.

On Academia/Industry collaboration Sandberg and Crnkovic [6] and Guillot et al. [8] investigate challenges between those parties and how to solve them with agile methods. Both analyse the adaption of the rather strict scrum methodology on said collaboration in multiple case studies with positive results. However, also that approach is highly dependent on team agreement.

Koetter et al. analyse the characteristics and problems of software development in distributed teams in research projects [10]. They give a literature review of common problems and typical solutions. With the focus on Software Architecture, the authors summarise the issues and sketch solution approaches on a methodological level.

3. Background: Continuous Integration

This section gives an introduction into the concepts of Continuous Integration. The next sub-section defines the scope of the methodology and gives a definition. Later sub-sections introduce the general concept and the Continuous Integration loop in more detail and introduce basics to testing and best practises.

3.1. Definition and scope

Continuous Integration describes a methodology to always have the latest successfully built and tested version of a software component available [2]. At its core, it aims at removing diverging developments of different developers by enforcing that all the code changes of every developer are integrated with each other to a shared mainline "all the time" (hence, it focuses on the integration of code of a single build artefact). Integrating small changes at high frequency reduces the chance of diverging code and the pain of code integration.

In Continuous Integration, the process of building and testing the component is usually described by scripts and hence, easy to reproduce by any developer and easy to automate. In consequence, it overcomes the issue of hard-to-reproduce builds that is a reoccurring problem in traditional development environments where developers usually have their code being built and run inside their different IDE in terms of version or even brand.

3.2. The integration pipeline

A successful adoption of Continuous Integration in any environment has to rely on automation in order to achieve a permanent feedback loop. This is illustrated in **Figure 1**. At some point in

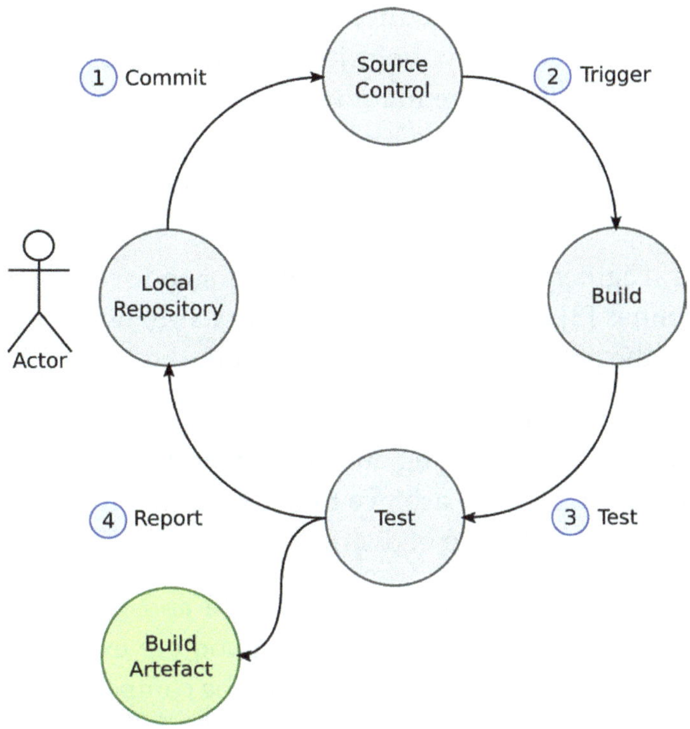

Figure 1. The Continuous Integration feedback loop realising the integration pipeline.

time, developers working on a local version of the code will be finished with their work, for example, a new feature or a bug fix. Then, they commit (step i) their changes to the version control system shared by all developers of that component. In addition to the code, the repository contains additional data and procedures to build and also test the software.

Accordingly, a new commit triggers (step ii) a new build of the software component. In case the build is successful (step iii), tests of the code get executed. Here, build automation enables that both building and testing can run automatically and do not require any human integration.

On a technical level, both build step and test step are executed on one or multiple build servers which is tightly integrated with the code repository and gets triggered through changes to the codebase. At the end of the build and test process, it will (step iv) report the status back to the users. Such a report includes information about failed builds or failed test cases. On success the build server issues a versioned and downloadable build artefact.

For closing the Continuous Integration loop, other developers react on reports issued by the build server. In case of successful build and test steps, they are supposed to immediately integrate the changes in their own code base. This core concept behind Continuous Integration ensures that the code bases of different developers evolve compatibly.

In order to successfully implement this feedback loop, it is important for every developer to commit very often (commonly interpreted as at least once a day). This ensures that merge conflicts stay minor and are easier to resolve.

3.3. Testing

Testing is necessary, as a successful build process does not give any hints whether the code is actually working. Hence, testing increases confidence on the codebase which creates an experiment-happy environment and reduces the risk introduced by possible ambiguity of requirements. Further, testing may yield information about code quality and runtime behaviour. Consequently, testing is the main vehicle to ensure reliability of and trust in the code. Obviously, this trust is higher, the higher the test coverage. Due to the many builds per day, testing can only be realised in an automated manner. In consequence, high automated test coverage is a core demand for Continuous Integration [2].

While there is no general agreement on a fixed number for code coverage percentage, there are suggestions and guidelines [3] about that metric. In practise, however, the desired coverage degree is dependent on the project and the criticality of the code.

As with the whole Continuous Integration methodology, it is important that all team members have understood the importance of testing and practise it. Unit and Integration Tests are the minimum amount of tests necessary to achieve that. Consequently, they are our main concern in this chapter. Further details on testing of distributed applications are available elsewhere [1].

Unit tests target small portions of code in the codebase and usually operate on a class- or routine-level. They are built alongside the application and are executed on a successful built. Implementing them makes sure that individual parts of the component are working as expected and intended. In contrast, integration tests are run against a fully built and unit-tested component. An integration test executes the software component as a whole and runs tests against APIs and if necessary utilises mocks.

3.4. Best practices

While the Continuous Integration loop as detailed earlier is simple, a true realisation of the approach requires flanking measures on the management and organisational side.

In order to decrease the change of a broken build and failing tests, developers have to build and test the application locally before committing their changes to the shared code repository. This practise leads to the desire that the automated test environment used on the build server and the test environment provided by the developer's IDE be compatible. Only then can the same tests be run in both environments and only then is the effort for the developer minimal to follow the principle of Continuous Integration.

Having such a set-up, a developer will commit more often to the shared code repository when experiencing short feedback cycles from the Continuous Integration loop. Ideally, the time from committing code changes to a tested software artefact is as short as running the tests on the development machine or even shorter.

However, even performing local tests will not avoid that at some point a build or a test will fail leading to a broken build. While the build is broken no developer should commit to the repository. Instead, everyone in the team is encouraged to contribute to fixing whatever caused the build to break. Only then further commits to the code repository are allowed.

In order to enable developers to witness that a build breaks and to trigger the process of handling this broken build, visibility of the current build and test status is a major concern. This can be as simple as a red or green badge being shown in a dashboard, an e-mail, but could also include bots that report on it on a messaging platform.

3.5. Summary

Summarising, Continuous Integration gives reproducible builds, versioned downloadable artefacts, and quick feedback on broken builds. Hence, it addresses many of the requirements brought up in Section 2: breaking down development into small units that are independently built and tested distributes work among partners (**R.1**) and at the same time, identifies responsible persons (**R.3**). Automating the build and testing process tremendously reduces the manual burden (**R.2**). It also checks dependencies on build level (**R.5**) and makes the software status visible (**R.6**). The availability of ready-to-use binaries is a first step towards an easy-to-use prototype (**R.4**). The definition of unit tests and integration tests allow people joining the project late to confidently make changes to the source code (**R.B**). If used properly, tests also serve as a testimonial of the currently defined requirements of the project (**R.C**). The separation of build and test phase enables some support for closed source code (**R.A**).

On the downside, Continuous Integration does not address dependencies on service level (**R.5**) and neither allows for a full easy-to-use set-up (**R.4**). In consequence, it also does not make the full software status available (**R.6**). With respect to closed and unavailable source code (**R.A**), further means have to be established.

Yet, the use of Continuous Integration also introduces new requirements:

R.CI.1 (additional project infrastructure): The use of Continuous Integration requires more infrastructure to be brought into the project. These include a revision control system, a build server, and a test server. All of them have to be maintained and explained to the consortium, for example, through tutorials. The build server in addition has to support all programming languages used in the project (**R.E**).

R.CI.2 (support for private code): For those partners in the project that want to disclose their source code to the public, the project infrastructure needs to support a private repository.

R.CI.3 (support for closed code): For those partners in the project that want to disclose the source code of their components even to the project, additional mechanisms have to be established in order to connect these components to the overall application.

R.CI.4 (team agreement): For Continuous Integration to work properly, all project partners have to agree on its use and be willing to take their share of the load. This is a management issue that can be supported when lean technology and good documentation is applied.

4. Background: Continuous Delivery

This section details background on Continuous Delivery. It starts with a definition and the usage scope, then presents the delivery pipeline and further testing steps. In contrast to Continuous Integration that has many challenges on social level, but a clearly defined build artefact at the end of a pipeline, the exact result of a run of the delivery pipeline is a design choice; the only demand is that it packages the binaries into something executable. Section 4.3 is concerned with the various possible approaches to packaging. Finally, when executing a component, various parameters may need to be configured, depending on the context the component is used in. We sketch possible design choices in Section 4.4.

4.1. Definition and scope

Continuous Delivery takes an executable binary (e.g. a build artefact) and packages it in a ready-to-run execution environment that resolves all internal and external dependencies, for example, to the operation system kernel, third-party libraries, and remote services. At this end, this automatic process creates packaged runtime environments for binaries and other artefacts. The rational is that pre-configured and tested self-contained packages are easy to roll out in different environments increasing the reliability of the roll-out process. In addition, abandoning manual actions strengthens maintainability and trust in the process.

When combined with Continuous Integration, Continuous Delivery provides a methodology that ensures that at any time a packaged, tested, and reliably deployable artefact is available based on the latest successful run of the integration pipeline.

4.2. The delivery pipeline

The delivery pipeline starts where Continuous Integration ends. It introduces the packaging step plus further automated and manual acceptance tests. A visual example of such a pipeline is shown in **Figure 2**.

Continuous Delivery starts with **build artefact(s)** that could be the outcome of Continuous Integration. The packaging step integrates one or more of them with any external dependencies and bundles them into packed artefacts (or simply artefacts). The pipeline is not necessarily linear and hence, can general more than one package. Depending on the process, packages for multiple architectures or use cases can be generated. While Continuous Integration contains

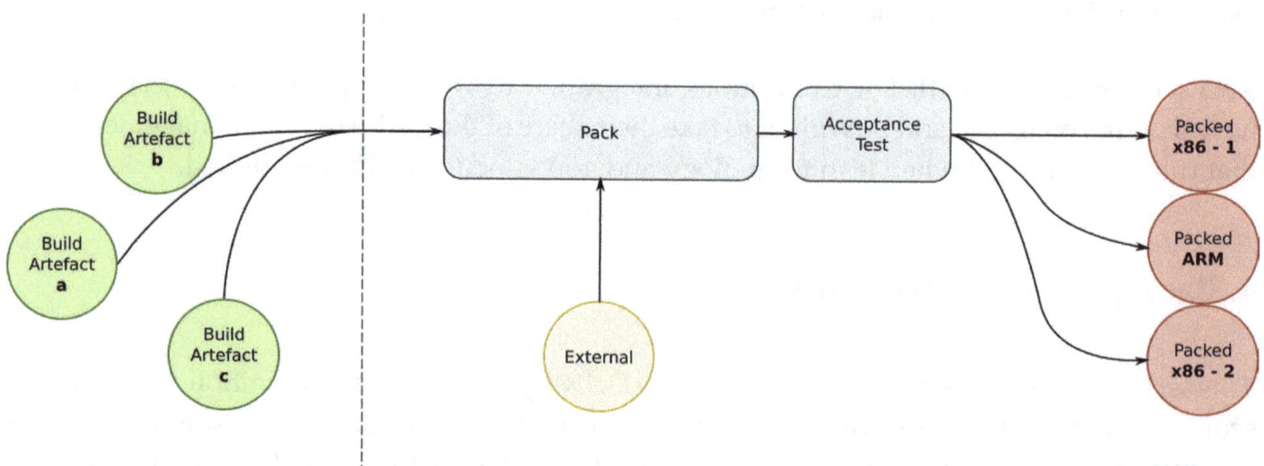

Figure 2. Example delivery pipeline.

a set of basic tests, Continuous Delivery introduces more sophisticated acceptance test. These are a crucial part of any useful delivery strategy and often contain both manual and automatic steps that validate if the software component behaves as expected. It will almost always include mocking of remote services.

4.3. Possible packing formats

The outcome of a run of the delivery pipeline is at least one deployable artefact packing an application component. While the delivery concept *per se* does not foresee a specific format and basically an arbitrary number of formats are possible, the following four approaches have found wide-spread acceptance and are commonly used. They differ in the size of the package, the coupling between component and host, and possible interferences with other components on the same host.

Virtual machine images package the component together with a suited operating system and all required third-party libraries. Executing the image in a virtual machine on a hypervisor introduces very strong runtime isolation between component instance (inside the virtual machine), the host installation (outside the virtual machine), and the host's hardware. It also creates barely any external dependencies and no direct interferences with other components on the same host. On the downside, virtual machine images are heavy weight in terms of size and resource usage. Container images are a lightweight alternative that also bundle the component with all third-party libraries. Yet, the container's operating system strongly depends on the hosting environment in terms of operating system version and kernel configuration. Still isolation between co-located components is available.

Both virtual machine and container images create an isolated and fully self-contained environment for the component. A conceptual different approach is followed by configuration management tools and distribution packages. Both of them install software directly on the host platform and barely create any isolation between different components. Software distribution packages are special archives that wrap the binary and files it ships with, but also contains hints to packages this binary depends on. Obviously, they integrate deeply with the dependency management of the host platform and utilise shared system resources and libraries directly. Configuration management tools provide a layer of abstraction, as they attempt to (re-)configure and change the hosts environment to reach a state which ensures that the application can run. They may do so by using software distribution packages. Both approaches are rather lightweight in terms of storage size.

4.4. Package configuration

The major goal of Continuous Delivery is to always have the latest deployable package of a component available. In consequence, this means that when creating the package, it is not known in which environment it will run. For instance, IP addresses, port numbers, paths to files may not have been defined yet, or can change over time. Consequently, when preparing a component for Continuous Deployment, it is important to foresee a configuration interface. Several approaches exist ranging from environment variables as suggested by the

12FactorApp[3] over configuration files as commonly used for components provided as Linux packages, to key/value stores or a database.

The choice of a configuration approach has influence on the overall implementation of a component. In addition, there is a mutual influence between configuration and packaging format.

4.5. Summary

Summarising, Continuous Delivery gives tested, versioned, and downloadable artefacts that are shippable, installable, and configurable out-of-the-box. As with Continuous Integration, the high degree of automation reduces manual burden (**R.2**). The use of Continuous Deployment helps the installation and management of the project outcomes (**R.4**) and at the same time helps remembering the big picture due to acceptance tests (**R.5**). The latter also contributes to the visibility of the software status (**R.6**). Similarly, acceptance tests help keeping track of desired outcomes (**R.C**) and support the fluctuation of team members (**R.B**).

On the downside, creating a larger deployable component from smaller parts, may counteract the equal distribution of load over partners (**R.1**) and makes the naming of responsible persons harder (**R.3**). Yet, when Continuous Integration is used in addition, logical bugs should have been filtered out and only those produced by acceptance test remain.

Continuous Delivery introduces the following additional requirements towards prototype integration and management.

R.CDel.1 (packaging format): Continuous Delivery requires to decide on one or more packaging formats per delivery pipeline.

R.CDel.2 (configuration options): Continuous Delivery requires to decide on the approach taken towards configuration per pipeline. It has to be consistent with the packaging format.

R.CDel.3 (support for closed artefacts): As with Continuous Integration additional mechanisms have to be established for any kind of closed code or closed binaries.

5. Background: Continuous Deployment

This section gives background on Continuous Deployment. As before, we start with a definition and set the scope for this methodology. Then, we describe the deployment pipeline. Finally, we consider deployment environments and application state.

5.1. Definition and scope

Deployment as such describes the process of enacting an application or application component. In general, it covers the steps from acquiring the necessary and possibly distributed hardware resources over installing as well as configuring the component(s) on these resources

[3]https://12factor.net/de/config

and starting the necessary deliveries. The task of deciding in what order components should be started is referred to as *orchestration*, the task of enacting components to find each other is called *discovery* or *wiring*.

Continuous Deployment describes a methodology to always have the latest version of all artefacts of an application deployed; and that updates to the application are visible in the deployment shortly after changes to the codebase. As with integration and delivery pipelines, the deployment pipeline is supposed to run automatically.

As all artefacts have gone through unit, integration, and acceptance tests, there is trust that individual artefacts work as expected. What is less reliable is the interplay of the components on an application-wide level. For that reason, in practise, multiple isolated environments are used and Continuous Deployment usually tackles the least critical environment, which is not linked with production systems. Yet, some companies like Amazon and Netflix demonstrate Continuous Deployment can go directly to production.

5.2. The deployment pipeline

The safety net of having multiple environments caters for incompatible version and interface changes of individual components. **Figure 3** shows an example of three traditionally used different environments as well as transitions between them.

The development environment contains the very latest version of the components' code and is automatically updated on every commit. In contrast, the production environment contains

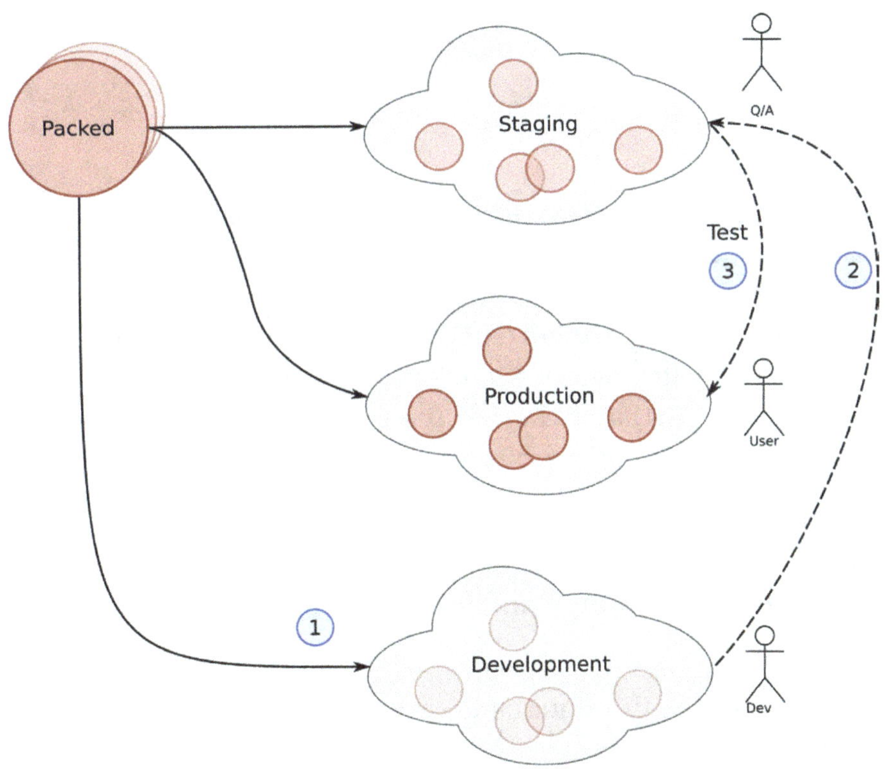

Figure 3. Deployment pipeline.

the actual live and fully functional environment facing users and customers. The staging environment is applied to validate updating the production environment to newer version. Therefore, staging uses a snapshot of the production data.

It is important to note that besides their build versions, the packaged components do not differ from environment to environment. What differs is their configuration in the respective environment (cf. Section 4.4) and the process of updating them. The development environment is automatically installed from scratch with each new deployable artefact from Continuous Integration. This environment is then used by developers in order to test and validate the common application. It is also used for reviews by Q/A. If these are successful, the version of the development environment is instantiated in the staging environment by updating the previous installation. This serves as a blueprint for updating the production environment. In case it succeeds, Q/A will enact more tests and finally decide to upgrade the production environment.

5.3. Application state

Usually at least one of the application components makes use of persistent state such as data stored in a database and on the file system. In order to support automatic re-deployment in case of failures and a seamlessly upgrade from one version to another, this state has to be separated from the artefact produced by the delivery pipeline. Otherwise, software and state cannot not be upgraded separately.

In consequence, state needs to remain available, even if the application environment is torn down. In IaaS clouds or containers, this can be achieved through the use of block storage/ volumes; in more traditional approaches, a remote file system, a NAS, or a SAN could be used. In consequence, the location of the state has to be configurable.

5.4. Summary

Summarising, Continuous Deployment realises support for a constantly deployed instance of the project outcome. In addition to that, it enables the realisation of use-case specific or demo-specific environment (**R.D**). Similar to Continuous Integration and Continuous Delivery, it helps distributing work among partners (**R.1**), reduces manual burden (**R.2**), and makes the software status visible (**R.6**). Its orchestration is the missing link to make available an easy to start and use environment (**R.4**) and clarifies the big picture (**R.5**). There are no immediate downsides to Continuous Deployment, but further requirements emerge:

R.CDep.1 (environment planning): The consortium has to agree on the number of environments and the desired flexibility in creating environments. In the most extreme cases the creation of a new environment is fully automatized and developers can flexibly create new environments.

R.CDep.2 (handling of state): Continuous Deployment requires to decide on the approach taken towards handling application state. In addition, stateful components need to be able to find their state, to validate it exists, and to initialise the storage location, in case it does not exists. In case state representation was changed, this has to be tolerated.

R.CDep.3 (support for closed artefacts): As before, additional mechanisms have to be established for any kind of closed binaries.

R.CDep.4 (additional infrastructure): In order to achieve the deployment and wiring of individual components, but also the whole project software, an orchestrator is necessary.

6. A research-oriented solution to software releases

In the following, we take the requirements put up in Section 2 and describe how we apply Continuous Integration, Continuous Delivery, and Continuous Deployment to support prototype integration as well as software releases for large-scale, widely distributed research projects. We also present how we address the additional requirements put up throughout Sections 3-5.

Section 6.1 presents the overall concepts and strategy we apply. The subsequent sections deal with the realisation of the individual pipelines. In each of these, we present our approach from a conceptual as well as a technical point of view and discuss the tools used. In addition, we present similar tools available on the market that could be used to provide the same or similar functionality.

6.1. Overview and concept

Sections 3–5 detail that the combined use of Continuous Integration, Continuous Delivery, and Continuous Deployment addresses almost all of the requirements established in Section 2. **Table 1** presents the coverage of requirements and methodology taken. Only the need to cater

	Continuous Integration	Continuous Delivery	Continuous Deployment
R.1	X		X
R.2	X	X	X
R.3	X		
R.4		X	X
R.5		X	X
R.6		X	X
R.A	(X)	(X)	(X)
R.B	X	X	
R.C	X	X	
R.D			X
R.E	X		

Table 1. Requirement mapping.

for closed code and binaries (**R.A**) is not naturally taken into account by any of the methodologies. It is, however, represented by the follow-up requirements R.CI.2, R.CI.3, R.CDel.3, R.CDep.3, and R.CDep.4 and has to be addressed by all three methodologies.

The following paragraphs sketch our approach on a high technical and management level.

6.1.1. General software set-up

Our approach centres around a project-wide code hosting platform that supports private repositories for code that shall not go public (**R.CI.3**). This platform is enhanced with a project-wide build and test server (**R.CI.1**) and further with an orchestration service linked to these two (**R.Dep.4**). Whenever possible, we rely on private hardware to host the needed infrastructures as well as the various environments of the deployment pipeline. In case this cannot be achieved, we fall back to a public cloud provider such as Amazon EC2 or Microsoft Azure.

6.1.2. Management process

In order to be able to apply Continuous Anything, decisions on the management level are required. These include first and foremost, the decision of the consortium to enact the methodology (**R.CI.4**). Once this decision is taken, the next step is to break down the overall project software into smaller components. This is a manual process that requires discussion and communication.

For each of the components an individual software repository is created and a responsible gets assigned. In an ideal case, exclusively members from one local team are responsible for one of these components. For each of the components then test cases are defined that detail how the component is supposed to interact with other components and more importantly that reflect requirements and goals of the project. Finally, an early integration pipeline for each component is realised that runs the tests. Only at this point, it is necessary that the developers of a component agree on a common technology including the programming language. Different components can agree on different languages.

In a next step, components are composed to deployable artefacts and for each of them a delivery pipeline is established. Acceptance tests are created and function as both a validation of the artefact's functionality as well as a representation of the project's requirements. Furthermore, a strategy towards packaging (**R.Del.1**) and configuration (**R.Del.2**) is decided upon. While it is principally possible to use different formats and approaches for different artefacts, the delivery pipeline benefits from unifying these.

In a last step, the consortium agrees on the number of environments to be used as well as the transition paths between environments (**R.CDep.1**). This is also the step where services with closed binaries get integrated in the whole system (**R.CDep.3**).

6.2. Continuous Integration

As clarified in Section 3, Continuous Integration requires additional infrastructure to be provided by the project. In particular, it demands for the operation of a code repository, a build server, and a test server.

6.2.1. *Concepts*

Due to the openness towards programming languages, the build server needs to cater for any reasonable programming language (**R.E**). We achieve that through specialised build environments. These build environments are used for the automated compiling and testing the components code (**R.2**) and are easily configurable and versioned (**R.B**) by the researchers themselves (**R.1, R.3**). The downloadable build artefacts are stored in the repository (**R.C**) with the appropriate access rights.

Access rights to each repository can be specifically set with permissions ranging from private (**R.CI.2**) over internal to public. Sometimes one (e.g. an industry partner) is not able to share their code or configuration parameters with the whole consortium or even publicly. Therefore, we limit access to code and encrypt certain configuration variables so only the actual owner has access to them (**R.CI.2**).

Regarding completely closed and private source code, which must not reside on the shared infrastructure (**R.CI.3, R.CDel.3**), we make the assumption that the build artefacts of these component are tested by their owners and their binaries available for use in the project. For closed binaries, we establish a customised deployment process (**R.CDep.3**).

6.2.2. *Selected tooling*

For our approach, we selected Git enhanced with GitLab (**R.CI.1**) as a source code repository. The primary reason for selecting this combination is due to state-of-the-art version control provided by Git as well as the user interface and eco-system provided by GitLab. Each software component is stored in an own Git repository.

As a build server we use GitLab Runner. On the one hand side this is due to its deep integration with GitLab, but on the other hand side, this is also due to its openness and flexibility also in supporting exotic demands (e.g. **R.E**). It achieves this, by enabling the use of custom build environments, giving the research teams a maximum amount of control.

Technically, each repository defines the build environment of the component stored in that repository. The environment also defines the integration pipeline and contains at least the two mandatory steps compiling and testing. When triggered, builds, and tests get executed in an instance of the defined build environment.

Due to the fact that we do not impose any programming languages, we do not rely on any specific build and dependency management frameworks. The same is true for testing frameworks. Here, the only requirement is that it can be included in the pipeline.

6.2.3. *Tooling alternatives*

The functionality we achieve through our set-up can also be realised through the use of other tools. For instance, Mercurial or SVN could be used as source code repository. Jenkins and Travis are alternatives for build servers.

With respect to build automation Maven is the de-facto standard for Java, while C programmers rely on make and pip could be used for Python. Testing can be implemented by JUnit or one of its derivatives for other languages.

6.3. Continuous Delivery

From Section 4, it is clear that the main challenge with Continuous Delivery are to decide on the packaging format and the configuration strategy.

6.3.1. Concept

For being able to easily start and use a component (**R.4**), we are packaging the artefact from Continuous Integration to make it executable. This process is automated by the build server (**R.2**). Once all the components from every partner have been packaged getting an instance of the application as a whole is comparatively low effort (**R.5**).

In contrast to the integration pipeline, not all repositories will have a delivery pipeline. Instead, multiple build artefacts can be combined to one packages artefact. For each packaged artefact, a root repository is selected that defines the delivery pipeline.

Similar to the integration pipeline, the process of packaging the component is specific to each repository. While the packaging format should usually be consistent for every component, it might be necessary to integrate with other build artefacts and external components, which is the task of the delivery pipeline.

While Continuous Anything does not demand for a specific packaging format on the concept level, the format should be (i) lightweight, to keep the delivery feedback cycle short, (ii) self-contained, to make acceptance testing easier, and (iii) configurable to cater for usage in different scenarios.

In order to support closed artefacts (**R.CDel.3**), we enhance the build server with a custom API that maintainers of closed artefacts are supposed to invoke (either automatically or manually) when a new version of their binary is available. This will then trigger the delivery pipeline for that artefact, if available or the delivery pipeline of artefacts that make use of it.

6.3.2. Selected tooling

Our approach does not impose any specific packaging format. Yet, for artefacts with standard demands, we encourage the use of Docker images, as they offer a good trade-off between isolation and ease of use. Containers are not as heavy weight as virtual machines, but still the software runs isolated. The resulting Docker images are pushed to an image repository, which is internal to GitLab for private artefacts or the public Docker hub for public ones.

For configuration, we suggest the use of environment variables for Docker containers as encouraged by good practises. For acceptance testing, we use the Selenium framework that enables record and playback of user interaction on interfaces.

6.3.3. Tooling alternatives

For packaging scenarios that demand higher isolation, virtual machines are the best choice. Here, Packer is a tool for the automated generation of virtual machine images. For configuration management, Puppet is an option, whereas for instance the Debian Package Manager can be used for creating distribution packages.

For configuration through key values stores, a myriad of different tools exists, ranging from Consul to classic databases, for example, MySQL, or even NoSQL databases, for example, MongoDB.

6.4. Continuous Deployment

Building on the decision we made in Section 6.3 by choosing Docker as packaging format, we need to align that to the additional requirements we set in Section 5. The following shows how we implement the Continuous Deployment of said containers.

6.4.1. Concept

We usually use the three basic environments **development**, **stating,** and **production** unless the project has special demands (**R.CDep.1**). Each repository with a delivery pipeline also has a deployment pipeline that automatically updates the development environment once a new packed artefact is available. Based on the development environment, the transitions between the other stages are handled as follows (**R.CDep.2**): (i) Upon manual decision, the artefacts deployed to development get redeployed in staging by overwriting the previous version. In addition, state from production is copied to staging for testing purposes. (ii) Going from staging to production is similar, except that no data are copied.

All environments are handled by an orchestrator and operated on a project-hosted infrastructure (**R.CDep.4**). For enabling state transition (**R.CDep.2**), this infrastructure comes with volumes to persist state and support mapping of state. For dealing with disclosed artefacts (**R.CDep.3**), we allow that callbacks are registered for each of them. These callbacks are used in order to trigger a new deployment or reset of the respective deployed artefact, as well as a transition of these deployed artefacts between their environments. The realisation of the callback is dependent on the responsible for the artefact. In addition, we introduce another API at the build server that owners of the closed artefact shall use to notify the environment about changes in their environment.

6.4.2. Selected tooling

The deployment in our system is done by the Rancher orchestration tool. For artefacts realised as containers, Rancher applies rancher-compose and docker-compose.yml files. These describe the actual configuration and a representation of the artefact to be deployed. Here, we can define the container (or virtual machine) image to use, the location of the state, and the desired configuration. Rancher also enables integrating external components.

6.4.3. Tooling alternatives

For orchestration of containers and virtual machines a plethora of different tools exist. These are either cloud-provider specific such as Amazon CloudFormation and OpenStack Heat, or reside outside the platform. In earlier work, we compare the features of these tools [19].

7. Discussions

Koetter et al. [10] are arguing that due to the tight schedule and different commitment of partners, a prototype integration is hard to achieve as it is too costly. This leads to only partially integrated systems that do not fully support all features. We argue that with our system, we have a clear and easy integration workflow that can be adopted by almost any (distributed) team with modern software development lifecycles and be adapted to existing ones. Therefore, we believe that our approach tackles the reported requirements and issues. Yet, our approach described in Section 6 is just one solution, from an overwhelming number of choices to make regarding the selection of tools and methods to realise Continuous Anything. The best possible technical realisation depends on what is currently used at the sides of the consortium members and familiarity of tools.

Nevertheless, team agreement as well as a clearly communicated and implemented methodology is more important than tool selection. The latter should always follow actual needs.

7.1. Project management culture

While Continuous Anything comes natural with the application of agile software development strategies, these are less an issue in distributed research projects. In this environment, it is hard to impossible to organise for instance daily stand up meetings or even weekly or bi-weekly sprints. As elaborated in Section 2 this is due to different schedules of the various stakeholders, the fact that barely anyone in the distributed team is dedicated full time to software development, and particularly that people travel a lot in order to promote the actual research work they do.

A possible way to work around the lack of a central pillar of the overall approach is to isolate responsibilities as much as possible and to only assign people from individual organisations to particular software components and have them organise the development process internally. This is the task of the project management.

A further core task of the project management besides organising the necessary infrastructure is to make sure that the integration strategy is rigorously followed from the beginning. This comprises the absence of shadow code, a common, shared understanding of how to use version control systems and when to apply changes to the master branch and other branches.

7.2. Software development culture

From all methodologies discussed, Continuous Integration can bring the most benefit. It is important to note, though, that everyone in the team has to agree on these practices being used. This might create some new pain points within the development team, but it is crucial that everyone understand the principles and share a common goal. This explicitly means that a non-trivial effort should be spent on test coverage. It is worth mentioning that for research-oriented environments daily commits of code are not necessary, but rather weekly or half-weekly commits are sufficient.

Continuous Delivery is especially helpful for research projects, since the described Big Bang Integrations usually happen multiple times during a project lifecycle; each time with a high risk of failing. The risk to failure puts a lot of stress on the whole consortium. In contrast, realising Continuous Delivery does not introduce new challenges except agreeing on a common packaging format.

Once Continuous Delivery has been realised, implementing Continuous Deployment is low effort. It is a crucial step to lower the effort for all consortium members to get access to a running instance of the project outcome.

8. Conclusions

In this chapter, we have presented our approach of Continuous Anything, a combination of Continuous Integration, Continuous Delivery, and Continuous Deployment in order to support the prototype integration to distributed research projects.

Our approach makes prototype integration a core element of the project plan and puts it on the same level as project management and financial administration. It does so by defining a framework that distributes and shares responsibility of integration work while at the same time clearly holding individuals responsible for dedicated software components. Through a high degree of automation, it keeps the overall integration work low, but still provides immediate feedback on the quality of the integration status. It is important to note that the quality of individual software components remains in the hands of their developers. It is them who decide which and if unit tests are necessary. In contrast, our framework requires that integration tests be available that ensure that interfaces between components work as intended. This approach allows an easy isolation of errors and the identification of responsible programmers in case of failures or problems.

Acknowledgements

This work has received funding from the European Union's Horizon 2020 research and innovation programme under grant agreements No. 732667 (RECAP), 732258 (CloudPerfect), and 644690 (CloudSocket).

Author details

Simon Volpert, Frank Griesinger and Jörg Domaschka*

*Address all correspondence to: joerg.domaschka@uni-ulm.de

Institute of Information Resource Management, Ulm University, Ulm, Germany

References

[1] Rother M. Toyota Kata. United States: McGraw-Hill Professional Publishing; 2009

[2] Duvall PM, Matyas S, Glover A. Continuous Integration: Improving Software Quality and Reducing Risk. Pearson Education; 2007

[3] Marick B. How to misuse code coverage. In: Proceedings of the 16th International Conference on Testing Computer Software; 1999. pp. 16-18. http://www.exampler.com/testing-com/writings/coverage.pdf

[4] Fowler M. Testing Strategies in a Microservice-Architecture [Internet]. Nov 18, 2014. Available from: https://martinfowler.com/articles/microservice-testing/ [Accessed: July 15, 2017]

[5] Fowler M. Continuous Integration [Internet]. May 1, 2006. Available from: https://www.martinfowler.com/articles/continuousIntegration.html [Accessed: July 15, 2017]

[6] Sandberg AB, Crnkovic I. Meeting industry: Academia research collaboration challenges with agile methodologies. In: Proceedings of the 39th International Conference on Software Engineering: Software Engineering in Practice Track. Piscataway, NJ, USA: IEEE Press; 2017

[7] Eddy BP et al. CDEP: Continuous delivery educational pipeline. In: Proceedings of the SouthEast Conference. New York, NY, USA: ACM; 2017

[8] Guillot I et al. Case studies of industry-academia research collaborations for software development with agile. In: CYTED-RITOS International Workshop on Groupware. Springer; 2017

[9] Fagerholm F et al. The RIGHT model for continuous experimentation. Journal of Systems and Software. 2017;**123**:292-305

[10] Koetter F, Kochanowski M, Maier F, Renner T. Together, yet apart – The research prototype architecture dilemma. CLOSER 2017 Proceedings of the 7th International Conference on Cloud Computing and Services Science. Porto, Portugal: SciTePress; April 24-26, 2017. pp. 646-653

[11] Eckstein J. Agile Software Development with Distributed Teams: Staying Agile in a Global World. United States: Addison-Wesley; 2013

[12] Sharma S, editor. The DevOps Adoption Playbook: A Guide to Adopting DevOps in a Multi-Speed IT Enterprise. United States: Wiley; 2017

[13] Humble J, Farley D. Continuous Delivery: Reliable Software Releases through Build, Test, and Deployment Automation. UK: Pearson Education; 2010

[14] Davis J, Daniels K. Effective DevOps: Building a Culture of Collaboration, Affinity, and Tooling at Scale. US: O'Reilly Media, Inc.; 2016

[15] Kim G et al. The DevOps Handbook: How to Create World-Class Agility, Reliability, and Security in Technology Organizations: IT Revolution Press; 2016

[16] Boehm BW. A spiral model of software development and enhancement. Computer. 1988; **21**(5):61-72

[17] Beck K et al. Manifesto for Agile Software Development [Internet]. 2001. Available from: http://agilemanifesto.org [Accessed: July 15, 2017]

[18] Booch G. Object oriented design with applications. Redwood City, CA, USA: Benjamin-Cummings Publishing Co., Inc.; 1991. ISBN: 0-8053-0091-0

[19] Baur D, Seybold D, Griesinger F, Tsitsipas A, Hauser CB, Domaschka J. Cloud orchestration features: Are tools fit for purpose? In: 8th International Conference on Utility and Cloud Computing. Piscataway, NJ, USA: IEEE Computer Society; 2015

9

Use of Renewable Energy for Electrification of Rural Community to Stop Migration of Youth from Rural Area to Urban

Urbanus F Melkior, Josef Tlustý and Zdeněk Müller

Abstract

Rural electrification is the key in developing countries to encourage youth and skilled personnel to stay in rural for production activities. Lack of grid network in Tanzania currently discourages youth and skilled personnel to live in rural areas. Tanzania has diverse renewable energy which needs to be developed for electricity generation. Most of these sources are found in rural areas but they are not developed and grid network are not extended because of low population density. The government has put in place policy which encourages small power producers to develop renewable energy resources. Energy produced would be sold to the community directly or to the government owned company for grid integration. This paper discussed three major renewable energy sources such as wind, solar and hydro power. Electrifying rural areas will encourage youth to reside in their communities and engaging themselves in production activities like farming and livestock keeping. Also communication among communities and networks between rural - urban would be improved. Establishment of small industries would lead more farm products and earn more money. Therefore, the strong links between rural - urban communities would be strengthened; hence youth migration would be stopped naturally.

Keywords: wind, solar, hydropower for rural electrification

1. Introduction

Tanzania is one among the few countries blessed with diverse primary source of energy, some developed and most of them undeveloped. The primary sources of energy available in Tanzania include hydropower (big, mini, micro, and pico), geothermal, solar, wind, biomass, coal, natural gas, and biogas. The primary energy used in the country is as follows: biomass (90%); petroleum products (8%); electricity (1.5%), and the remaining (0.5%) contributed by coal and other renewable energy sources [1]. However, more than 80% of energy obtained from biomass is consumed in rural areas, hence contributing to deforestation as well as damaging rural people's health since they inhale smoke from wood when preparing food for the family. Electricity network did not cover the rural area because extending the grid to rural areas is not financially and economically feasible. This situation forced youth people to migrate from the rural areas to urban areas where there is reliable electricity for various income generating and social activities.

2. Renewable energy in Tanzania

Tanzania has plenty of renewable energy sources of which few of them are developed for electricity generation. Currently, a large-scale hydropower resource has been developed for electricity generation, while the small hydropower, which has good potential and is particularly feasible in rural areas, is not developed for electricity generation.

Furthermore, biomass resources are mostly exploited in unsustainable ways resulting in cutting of trees, hence causing the environmental degradation. The country has great potential of organic waste generated from the agricultural sector and remains unexploited.

Tanzania has enough solar insolation which is suitable for off-grid and grid as well. The solar insolation is high in the central part of the country. Currently, the solar energy resources are utilized in small scale particularly for roof-mounted panels for powering domestic appliances such as lighting and mobile phone charging.

Furthermore, wind energy resource assessment in the country indicates its viability; therefore, the plan for developing wind farm is in progress.

This chapter will discuss the availability of renewable energy and the calculation/formula for estimating available energy contained in the energy source.

2.1. Wind energy

2.1.1. Introduction

Wind is a widely distributed energy source, between 30°N and 30°S. Earth is unevenly heated by the Sun resulting in the poles receiving less energy from the Sun than the Equator does, also dry land heats up more quickly than the seas do. This difference in heating gives power to a global atmospheric convection system reaching from the Earth's surface to the stratosphere which acts as a virtual ceiling. Heated air at the Equator rises and is replaced by cooler air coming from the south and the north [2]. That is, cool winds blow toward the Equator. Tanzania is situated near the Equator; it is affected by the air movement as well as benefits from this prevailing condition.

The availability of wind varies for different regions and locations. It has been noted that there is a period in a year that the wind speeds are higher, and some period, the wind speeds are low. Due to seasonal variations, the potential of wind for power generation can be significantly higher than the annual mean wind speed would indicate [3].

Thus, when embarking to the project, not only the mean wind speed but also the wind speed frequency distribution, commonly described by a Weibull distribution, has to be taken into account in order to estimate accurately the amount of electricity that would be generated [4]. Wind speed varies with height, depending on surface roughness and atmospheric conditions. Daily and hourly variations in the wind speed are also important for scheduling the operation of the conventional power plant and adjusting their output to meet these variations [5].

Based on the available information, Tanzania has plentiful wind resources, with much of it located around the Great Lakes, the plains, and the highland plateau regions of the Rift Valley [6].

The wind is the sustainable energy source which does not create emissions and it will never run out since it is constantly replenished by energy from the Sun. Generation of electrical power can be done by wind turbine which converts wind energy to mechanical power to drive an electrical generator. Wind turbines undergo natural evolution from traditional windmills with the several blades to three blades, which rotate around a horizontal hub at the top of a steel tower [7]. Wind passes over the blades exerting a rotating force to turn a shaft which

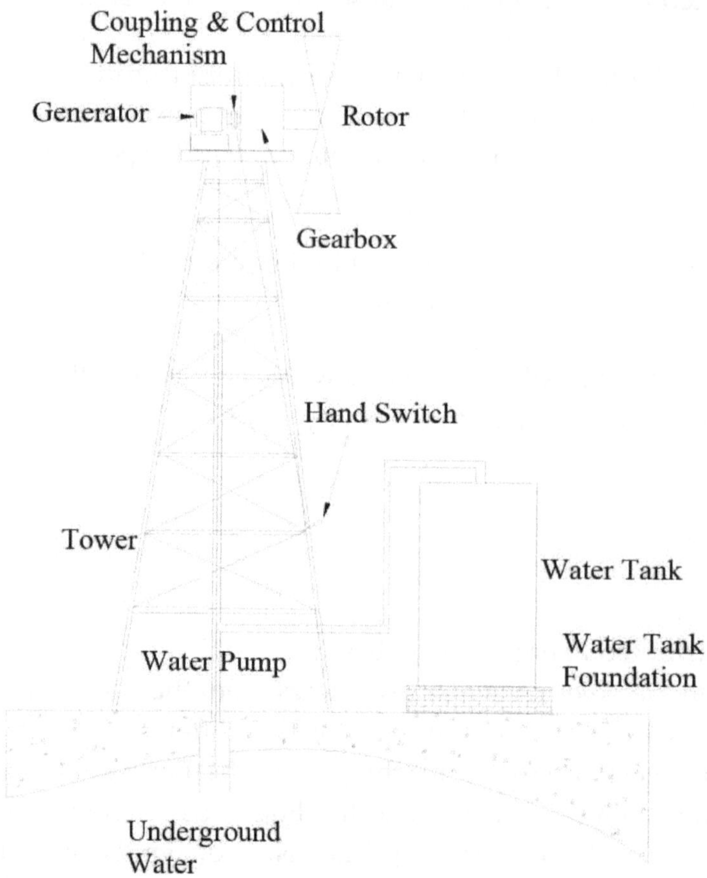

Figure 1. Dual purpose windmill for water pumping and electricity generation.

connects with gearbox enclosed in the nacelle. An electrical generator either induction or synchronous is connected to the gearbox which will amplify speed shaft obtained from the blades (rotor). Most wind turbines start generating electricity at wind speeds of around 3–4 m/s, maximum power would be generated at around 15 m/s and generation stops at the wind speed of 25 m/s by shutting down to prevent storm damage [8] (**Figure 1**).

Wind energy resource in Tanzania would be assessed in the different location of the country as follows: east zone (coastal area), central zone, west zone, southwest zone, southeast zone, northeast zone, and northwest zone (lake zone). Each zone has weather station equipped with measuring wind speed, humidity, temperature, and rainfall. The information collected from individual weather stations have always been submitted to Tanzania Metrological Agency for further processing.

For this study, monthly average wind speed data for 9 years (January 2009–July 2017) measured at the height of 3 m were collected from weather stations located strategically all over the country. The average wind speed shows that in the central zone there is reasonable wind speed which varies from 2.8 to 6.2 m/s; therefore, the place is suitable for the wind farm as well as isolated wind power plant (**Figure 2**). Average wind speed for the east zone found to be ranging from 3.4 to 5.8 m/s but the place is not suitable for the wind farm because of the scarcity of land but can be used for isolated power supply (**Figure 3**). In west zone, wind speed ranges from 2.5 to 4.2 m/s which is suitable for isolated power supply (**Figure 4**). In southwest zone, wind speed ranges from 2.2 to 4.8 m/s which is suitable for isolated power supply

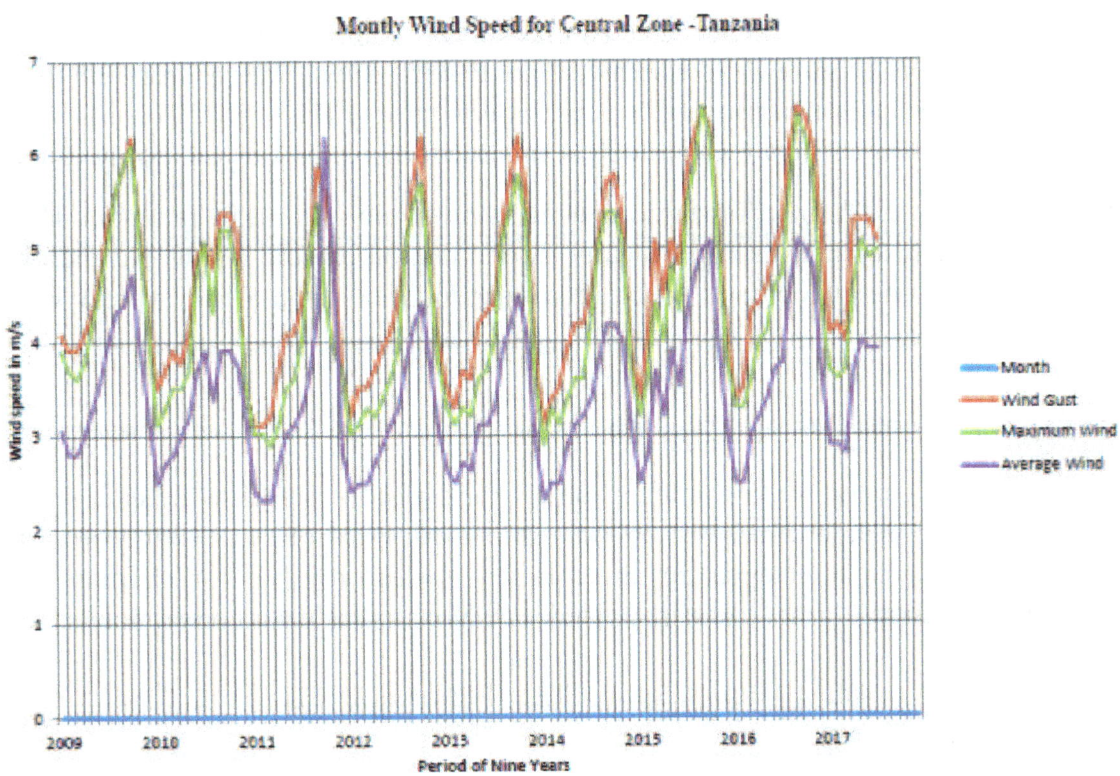

Figure 2. Average wind speed for central zone of Tanzania.

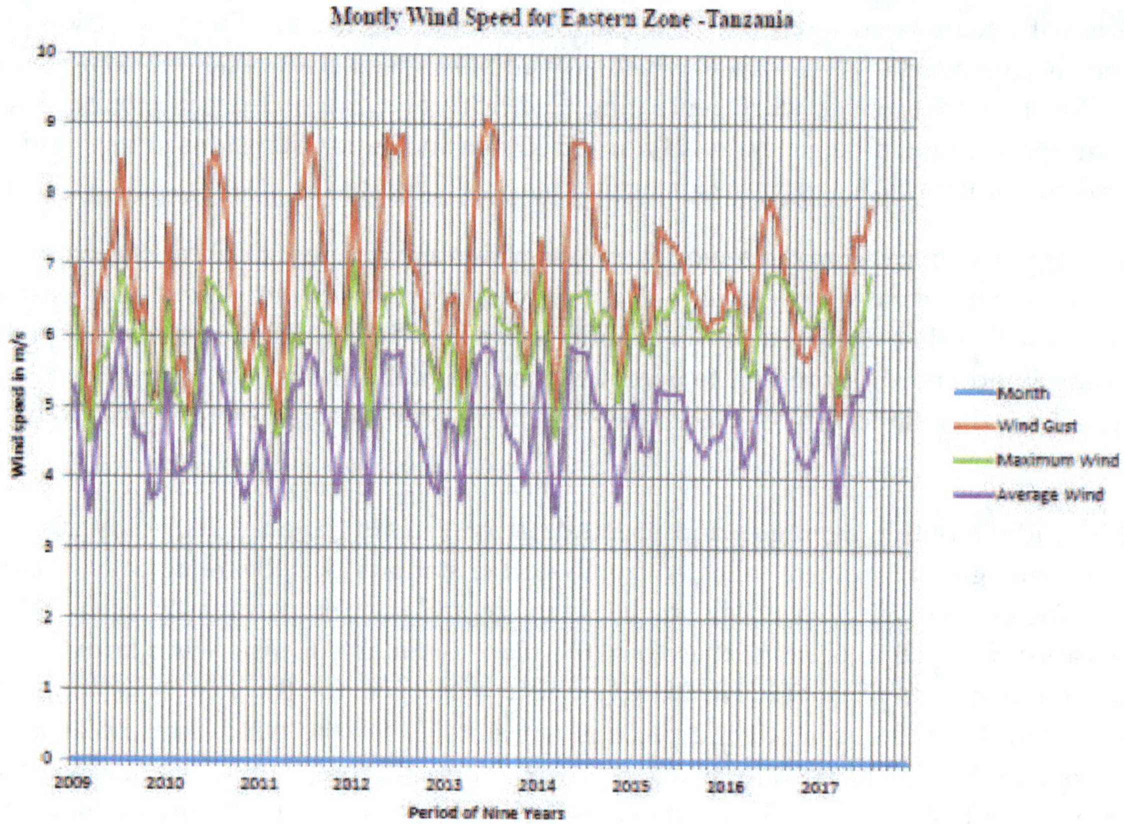

Figure 3. Average wind speed for eastern zone of Tanzania.

(**Figure 5**). In southwest zone, wind speed ranges from 3.8 to 6.7 m/s which is suitable for isolated power supply (**Figure 6**). In northwest zone, wind speed ranges from 2.8 to 4.2 m/s which is suitable for the wind farm and isolated power supply (**Figure 7**). In southwest zone, wind speed ranges from 2.7 to 4.4 m/s which is suitable for isolated power supply (**Figure 8**).

To date, there is no wind farm in Tanzania or even stand-alone systems in Tanzania. Therefore, the country waived taxes such as import, value-added tax to promote renewable energy in the country. Therefore, energy demand for grid connections or stand-alone system would be estimated.

2.1.2. *Theory and principles of wind energy conversion*

Having the energy demand for grid integration or stand-alone systems, wind energy parameters and equipment would be estimated. Wind turbines can extract kinetic energy from air that passes through the area intercepted by the rotating blades only [9]. For air mass m in kg moving at speed U in m/s, kinetic energy in Joules or Nm available for conversion is:

$$KE = \frac{1}{2} \times m \times U^2 \text{ or } KE = \frac{1}{2} \times \rho_a \times V \times U^2.$$

where

ρ_a = air density in kg/m^3, V = volume of air in m^3, m = mass in kg/s and can be expressed as:

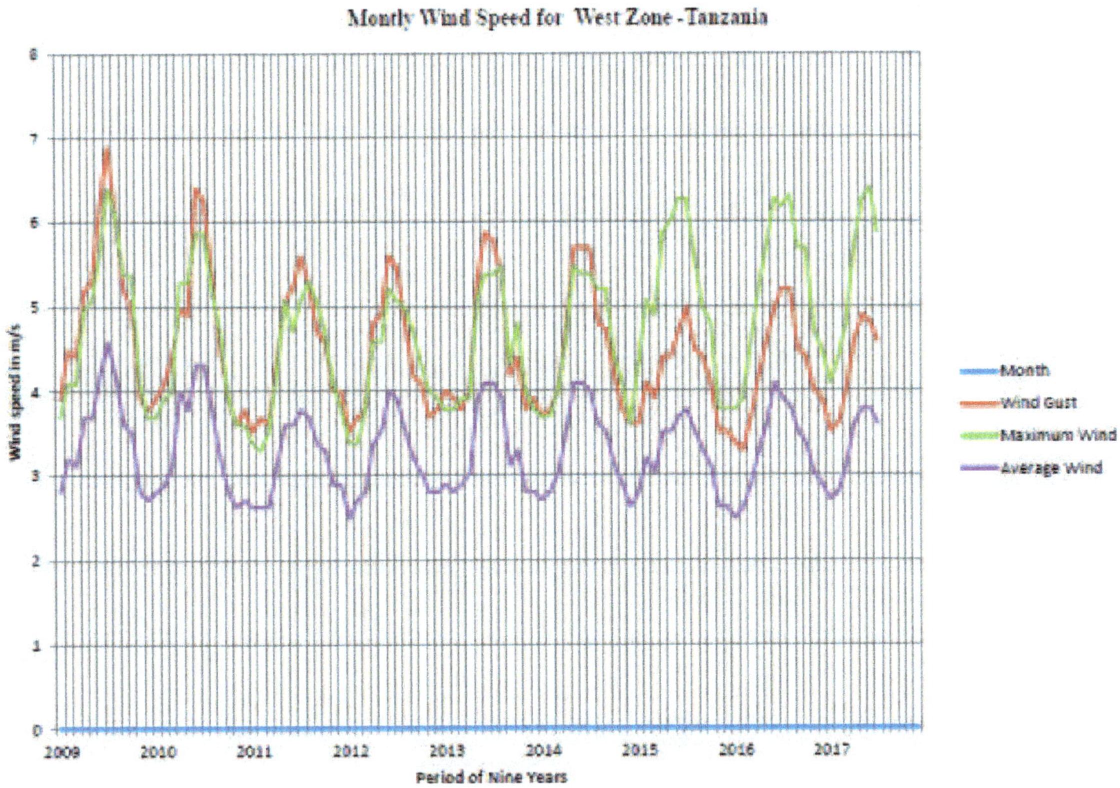

Figure 4. Average wind speed for western zone of Tanzania.

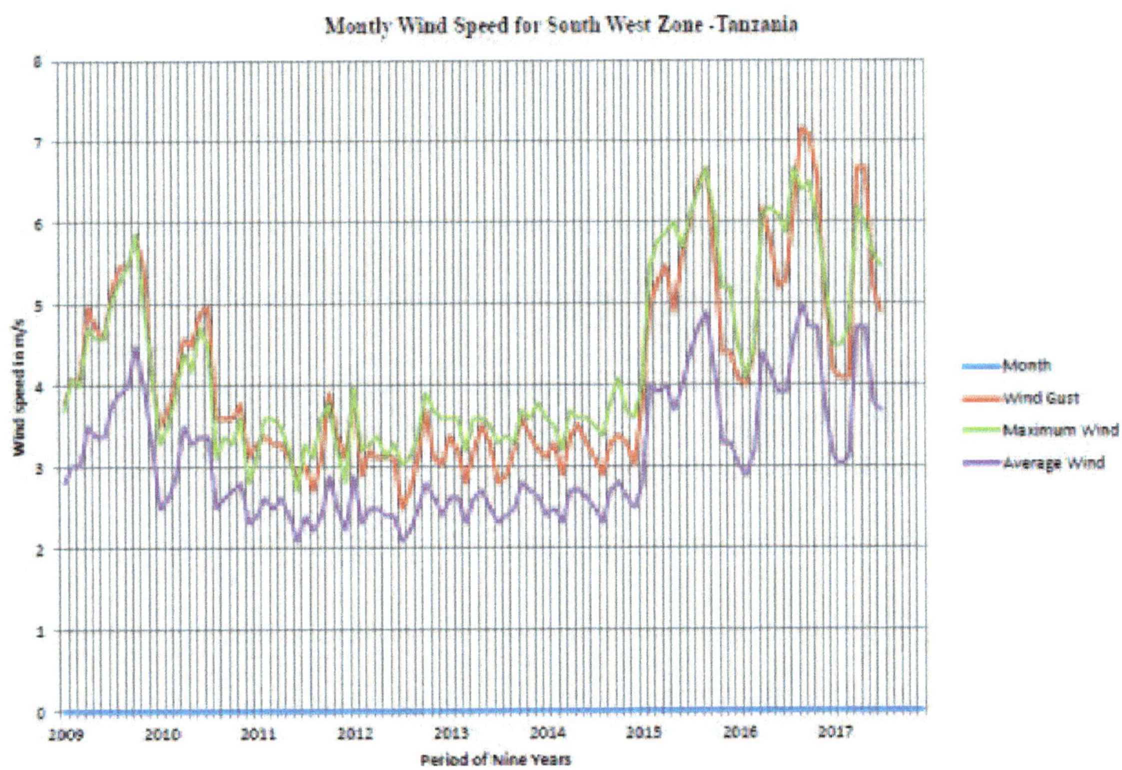

Figure 5. Average wind speed for southwest zone of Tanzania.

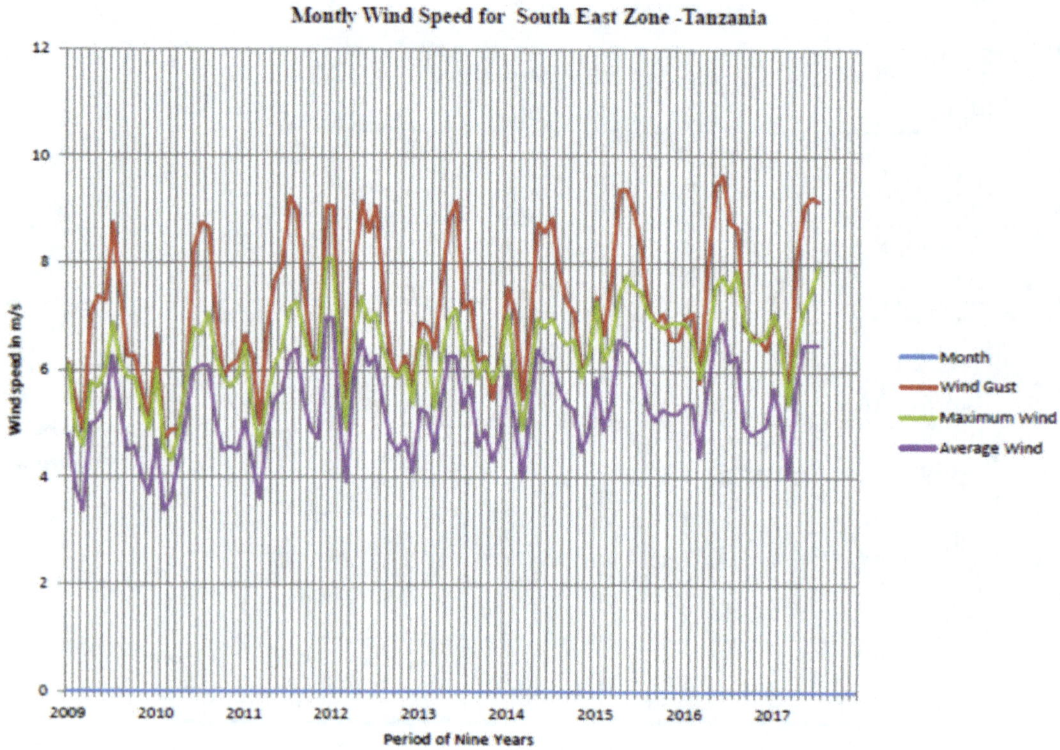

Figure 6. Average wind speed for southeast zone of Tanzania.

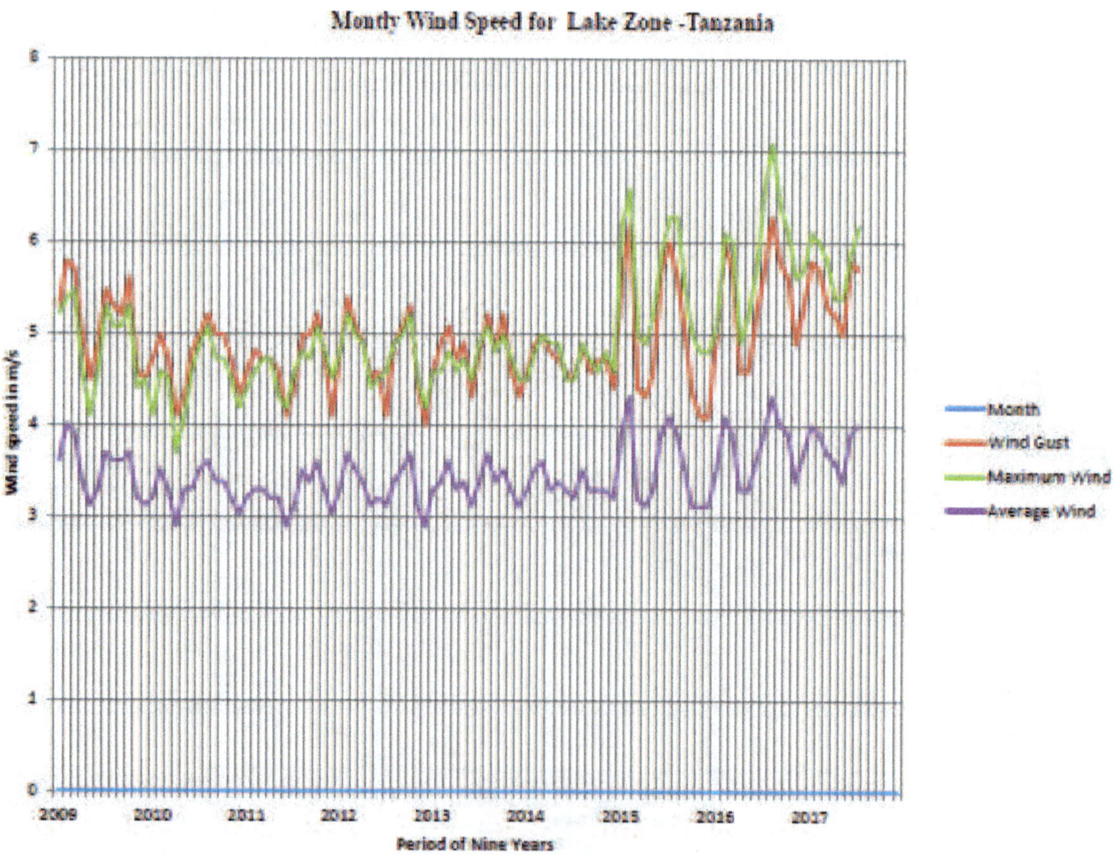

Figure 7. Average wind speed for lake zone (northwest) of Tanzania.

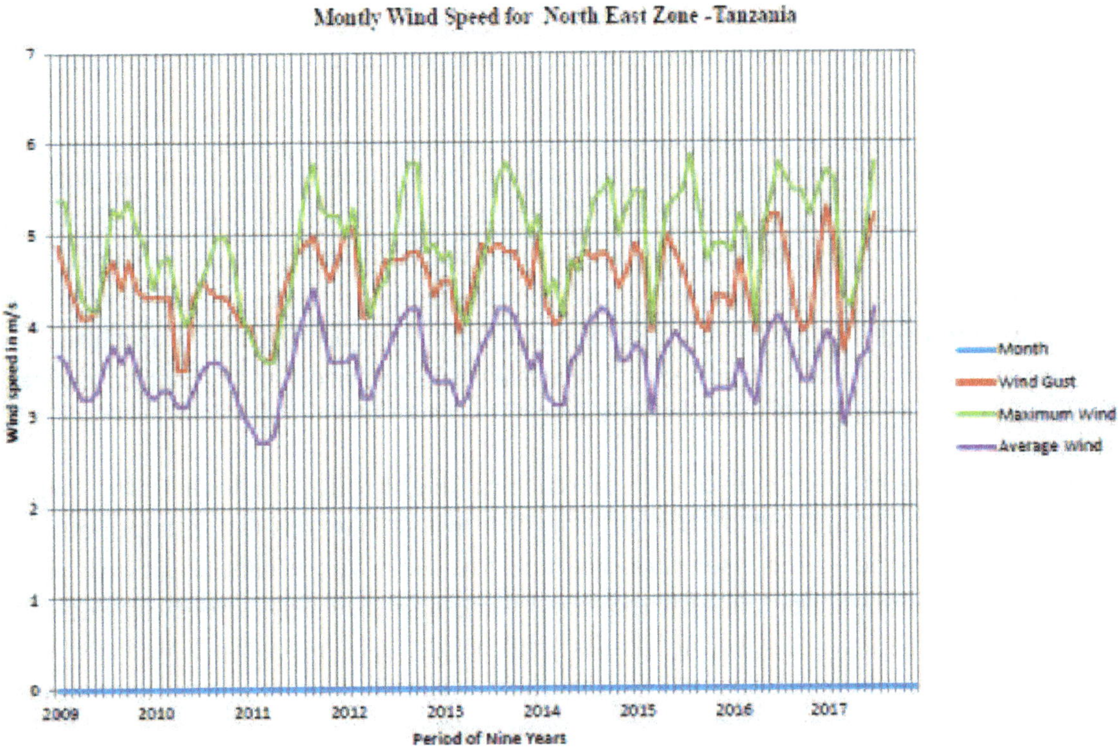

Figure 8. Average wind speed for northeast zone of Tanzania.

$$m = \rho_a \times A \times U$$

Then, available wind power P in watts is:

$$P = ½ \times \rho_a \times A \times U^3$$

Power density or power per unit area in W/m^2 of a particular site is:

$$\frac{P}{A} = ½ \times \rho_a \times U^3$$

Shaft power that can be obtained from a wind turbine in W is:

$$P = ½ \times \rho_a \times A \times U^3 \times C_p \times \eta_t$$

where P = shaft power in watts, A = swept area for the turbine in m^2, U = wind speed in m/s, ρ_a = air density in kg/m^3 (around 1.225 kg/m^3 for 15°C at sea level), η_t = turbine efficiency, and C_p = coefficient of performance that depends on rotor speed and wind speed, i.e., the tip-speed ratio.

Typically, C_p ranges around 0.5 for large electricity-generating wind turbines and 0.35 for water pumping wind turbines. Taking into account generator efficiency η_g, power output from electricity-generating wind turbines is:

$$P = ½ \times \rho_a \times A \times U^3 \times C_p \times \eta_t \times \eta_g.$$

2.1.3. *Speed extrapolation*

Wind speed measurement is usually done by using an anemometer. The anemometer has three cups and vane for capturing wind speed and detects its direction as well. The instrument is normally being installed in a location where wind is free with no influence from nearby object [10]. For the installation of this instrument, economical factor as well as the degree of accuracy is required (as the height increases, the cost of installation increases). Most of the anemometers are installed at the height of 3–10 m; hence, there is a need for speed extrapolation to determine the wind speed at the height of the wind turbine.

Typically, the increase in wind speed with increase in height follows a logarithmic profile that can be reasonably approximated by the wind profile power law, using an exponent of 1/7th, which predicts that wind speed rises proportionally to the seventh root of altitude. Doubling the altitude of a turbine will increase the expected wind speed by 10% and the expected power by 34%. In general, then, wind speed increases with height in some complicated and turbulent way depending on local conditions and topography. Nevertheless, two velocity extrapolation laws exist:

- the 'log law' and
- the 'power law'

These laws can be used to predict the wind speed at the height of power generation from wind speeds measured at a lower height.

2.1.3.1. *Log law*

The log law, which can be derived theoretically using several different methods, is given by:

$$U_Z = U_{Zr} \times \frac{\ln(Z/Z_o)}{\ln(Z_r/Z_o)}$$

where U_Z = speed at the height of power generation, U_{Zr} = speed at the height of measurement, and Zr and Zo = roughness length.

Roughness lengths range from 0.01 mm for wind flowing over smooth ice or mud to 10 mm over rough pasture to 0.5 m over forests and woodlands.

2.1.3.2. *Power law*

In power law equation, wind shear is quantified as the exponent n that relates wind speeds at two different heights. Wind shear is possible to be calculated when upper and lower wind speed measurements are available. Wind shear depends on the nature of land surface that is smooth or rough. The areas with trees and buildings will produce more friction and turbulence than smooth surfaces (lakes or open cropland). The greater friction means the wind speed near the ground is reduced. The approximate increase of speed with height for different surfaces can be calculated from the following equation:

The purely empirical power law is given by:

$$U_Z = U_{Zr} \times \left(\frac{Z}{Zr}\right)^n$$

where

n = exponent determining the wind change,

Zr = reference height,

Zo = roughness coefficient,

U_{zr} = wind speed at the measurement.

But, in practice, wind speed varies with elevation, time of day, season, topography, wind speed, temperature, and other factors. However, increasing more the height of turbine will bring the difficulties about the cost of erecting the tower as well as the cost of foundation. The different values of *n* were indicated in **Table 1** for the different ground natures.

S/N	Type of terrain	n
1.	Smooth sea or sand	0.10
2.	Low grass steppe	0.13
3.	High grass and small bushes	0.19
4.	Woodlands and urban areas	0.32

Table 1. Values of n for different ground covers.

2.2. Solar energy

2.2.1. Introduction

Sunlight is a general term for the electromagnetic radiation emitted by the Sun that can be collected and turned into useful forms of energy, such as heat and electricity, using various technologies. However, the technical feasibility and economical operation of these technologies at a specific location depends on the available solar resource [11].

The Sun rays strike the Earth at the angles ranging from 0° (just above the horizon) to 90° (directly overhead), because the Earth is round and it revolves the Sun and its orbit. The Earth surface will have maximum energy only when the Sun's rays are vertical, as the more Sun's rays are slanted, the long way they travel through the atmosphere, becoming more scattered and diffuse [12]. The Earth revolves around the Sun in an elliptical orbit and is closer to the Sun during part of the year.

When the Sun is nearer the Earth, the Earth's surface receives a little more solar energy. The Earth has great lines running from west-east namely Equator (0°), Tropic of Cancer (23.5°), north of the Equator (passes through Mexico, the Bahamas, Egypt, Saudi Arabia, and India), and Tropic of Capricorn (23.5°), south of the Equator (passes through Australia, Chile, southern Brazil, and northern South Africa).

Figure 9 shows the energy balance, as the Sun emits solar energy of 173 × 10^{12} W, at the atmospheric boundary, 30% will be reflected and 70% (121,000 × 10^{12} W) will reach the Earth. As the Sun comes to Earth, its 30% of energy would be absorbed by the atmosphere and remaining 70% is used in tidal, wind wave, evaporation, fossil fuel, hydro, geothermal, and photosynthesis.

Sunshine is part of the radiation that is supplied by the Sun, especially light, infrared, visible, and ultraviolet light. On Earth, the Sun is filtered through the atmosphere, and it is daylight when the Sun is on the horizon. When the direct sunlight is not limited to the clouds, it has as much sunlight as a mixture of bright light and heat. If it is blocked by clouds or resembles other objects, it is like a diminishing light. Therefore, sunshine is the most important factor for photosynthesis, the process used by plants for converting light energy to chemical energy.

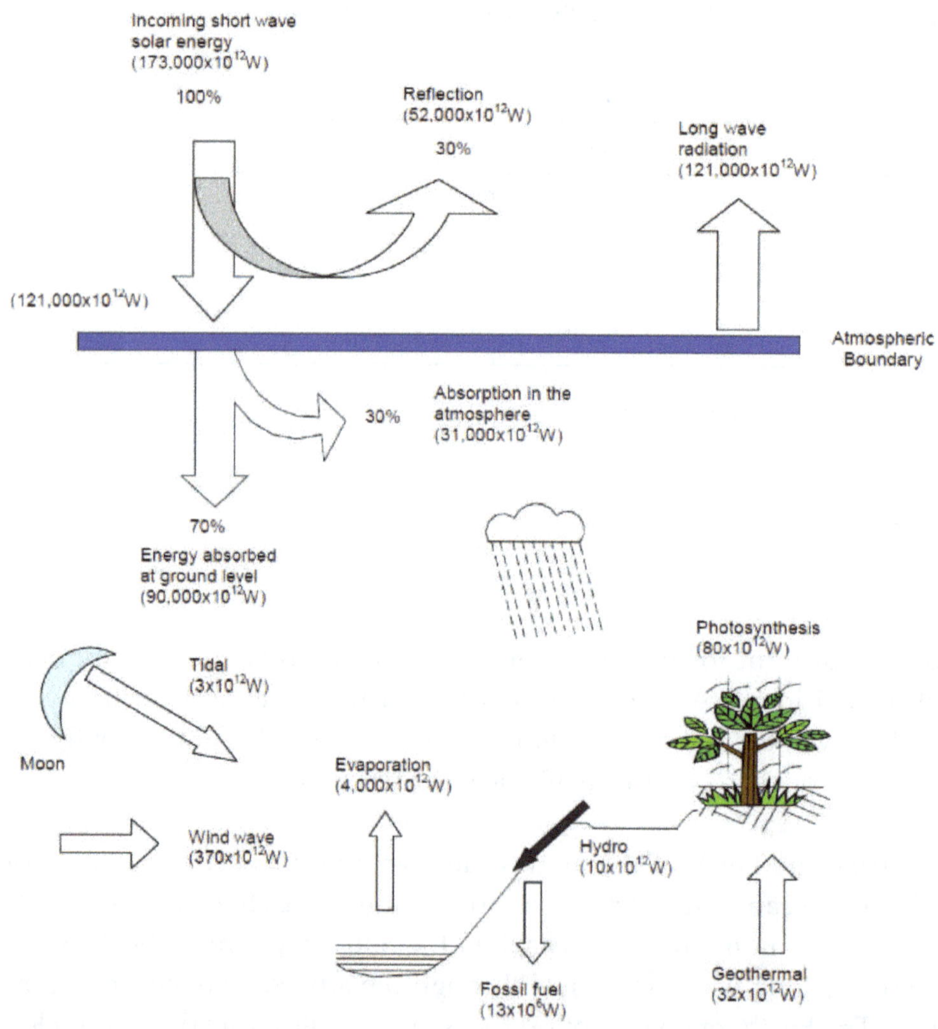

Figure 9. Energy balance.

During photosynthesis, the plant produces carbohydrates and oxygen to form water and carbon dioxide using sunlight as the source of energy [13].

The Sun is overhead at the Tropic of Cancer on June 21 which is the summer in the northern hemisphere and winter in the southern hemisphere. Also, on December 21, the Sun is overhead on the Tropic of Capricorn which is summer in the southern hemisphere and winter in the northern hemisphere. The area bounded by the tropics that are Tropic of Cancer on the north and Tropic of Capricorn on the south experiences tropical climate. The area under tropic climates experiences two seasons in a year that is rainy season and dry season. However, the area in the north of Tropic of Cancer and that at the south of Tropic of Capricorn experiences four seasons in a year that are winter, falls, spring, and summer [14].

The amount of solar radiant energy incident on a surface per unit area and per unit time is called irradiance or insolation. The average extraterrestrial irradiance or flux density at a mean Earth-Sun distance and normal to the solar beam is known as the solar constant, which is 1366.1 W/m^2 according to the most recent estimate [15]. The energy delivered by the Sun is intermittent and changes during the day and with the seasons. Photovoltaic (PV) conversion is the direct conversion of sunlight into electricity [16]. Photovoltaic devices are rugged and simple in design and require very little maintenance, constructed as stand-alone systems to give outputs from microwatts to megawatts and have been used as the power sources for calculators, watches, water pumping, remote buildings, communications, satellites and space vehicles, and even multimegawatt scale power plants. With such a vast array of applications, the demand for photovoltaics is increasing every year [17].

Solar energy exhibits the highest global potential since a number of factors such as latitude, diurnal variation, climate, and geographic variation are largely responsible for determining the intensity of the solar influx that passes through Earth's atmosphere. The average amount of solar energy received at Earth's atmosphere is around 342 W/m^2, of which 30% is scattered or reflected back to space, leaving 70% (239 W/m^2) available for harvesting and capture [18]. The annual effective solar irradiance varies from 60 to 250 W/m^2 worldwide.

Tanzania, an East Africa country, is situated just south of the Equator and is lying in between the area of the Great Lakes such as Lake Victoria, Lake Tanganyika, and Lake Nyasa and the Indian Ocean. It contains a total area of 945,087 km^2 including 59,050 km^2 of inland water. It is bounded on the north by Uganda and Kenya, on the East by the Indian Ocean, on the South by Mozambique and Malawi, on the southwest by Zambia, and on the West by Democratic Republic of Congo, Burundi, and Rwanda. Tanzania has a latitude and longitude reading of 6°00′ south and 35°00′ east.

Table 2 gives the insolation level values in some areas of the country captured by the study. Solar photovoltaic energy is uniquely useful in rural not served by the national grid to provide basic services such as irrigation, refrigeration, communication, and lighting. Solar energy is often more efficient than traditional sources such as kerosene. For lighting, a photovoltaic compact fluorescent light system is more efficient than kerosene lamp, used in rural areas to provide night lighting. Photovoltaic system also avoids high costs and pollution problem of standard fossil fuel power plant (**Figure 10**).

Zone	Solar insolation											
	Jan	Feb	Mar	Apr	May	Jun	Jul	Aug	Sep	Oct	Nov	Dec
Central	6.1	6.0	6.1	5.7	5.6	5.8	5.7	6.0	6.3	6.4	6.5	6.2
East	5.2	5.3	4.9	4.0	4.3	4.4	4.4	4.0	4.9	5.1	5.8	5.6
Southwest	6.0	6.1	5.7	5.9	6.2	6.3	6.1	6.6	6.7	7.0	6.7	6.2
West	4.3	4.5	4.9	4.3	4.4	4.8	4.3	4.9	4.9	4.7	4.1	4.3
Southeast	4.4	4.6	4.3	4.0	4.4	4.4	4.5	4.6	4.9	4.9	5.2	4.8
Northwest	5.4	5.0	5.4	5.4	5.4	5.0	5.2	5.4	5.4	5.4	5.7	5.4
Northeast	5.6	5.5	5.6	4.7	3.6	3.8	4.0	4.1	4.6	5.0	5.4	5.6

Table 2. Mean monthly solar insolation.

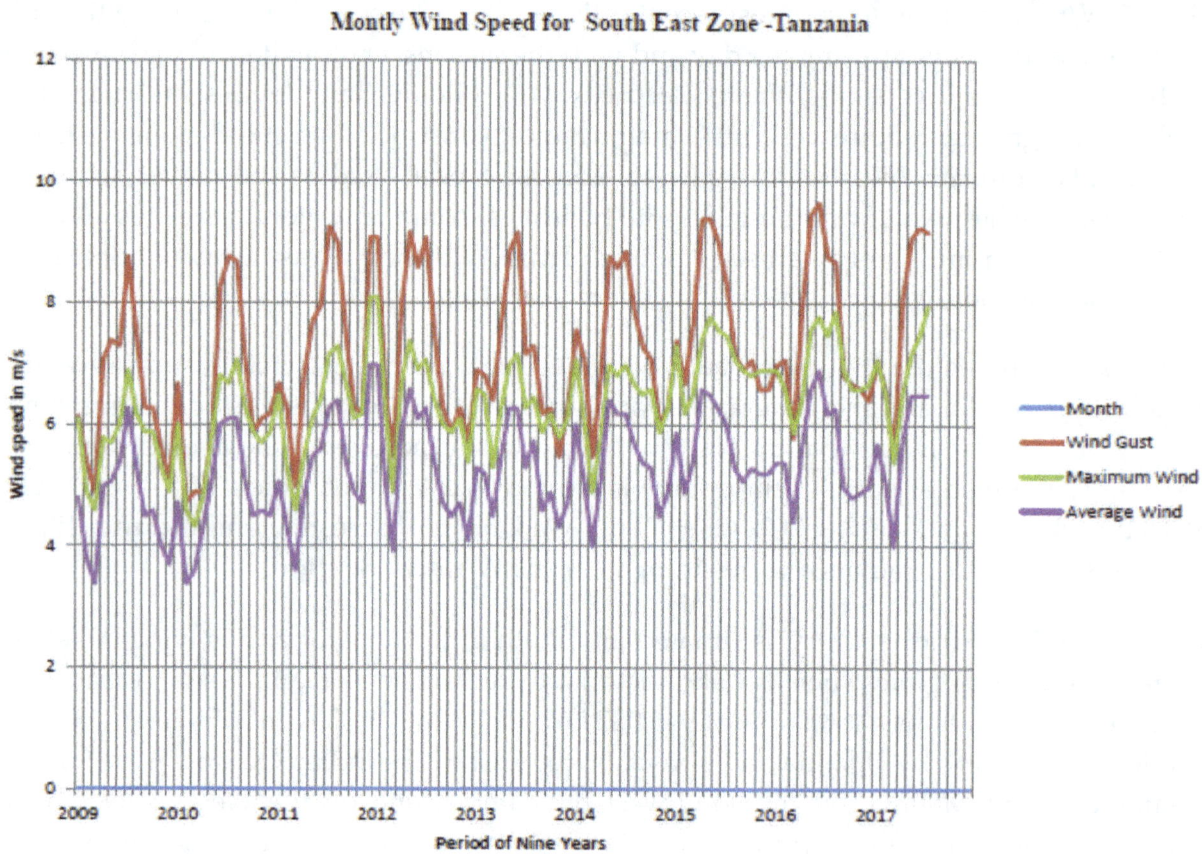

Figure 10. Solar insolation in Tanzania.

2.2.2. Solar PV system sizing

The electricity generated by solar photovoltaic can be stored or used directly, fed back into grid line or connected directly to customers. Solar PV system includes different components that should be selected according to your system type, site location, and applications. The solar PV system consists of the following components: solar panel, solar charge controller, inverter, battery bank, auxiliary energy sources, and loads (appliances).

2.2.3. *Determine power consumption demands*

Before embarking to system design, all appliances that need to be supplied by the solar PV system have to be identified.

- Record all power ratings of all appliances to be supplied by the solar PV;
- Estimate average hours need to operate each appliance;
- Calculate the energy need by multiplying appliance ratings with estimated hours of use per day;
- Add the watt-hours of all appliances together to get total watt-hours per day;
- Multiply the total appliance watt-hours per day by 1.3 (the energy lost in the systems such as panel, inverter, charger controller, battery, and wiring systems) to get the total watt-hours per day which must be provided by the solar panels.

2.2.4. *Size the PV modules*

Solar PV exists in different sizes and ratings. The size of solar PV modules depends on total peak watt-hour needed, the climate of the site location, and panel generation factor. The panel generation factors for different locations in Tanzania are given in **Table 2**. The sizing of PV modules is calculated as follows:

- Calculate the total watt-peak ratings for PV modules by taking total watt-hours needed per day divided by the panel generation factor given in **Table 1** to get the total watt-peak rating for the PV panels needed to operate the appliances;
- Calculate the number of PV panels for the system by dividing watt-peak ratings with the rated output watt-peak of the PV modules available. Increase any fractional part of result to the next highest full number and that will be the number of PV modules required.

2.2.5. *Inverter sizing*

An inverter is the power converter used to convert direct current power (DC) to alternating current power (AC). The inverter is needed when the available appliances need AC power. The size of the inverter should never be lower than the total watts of appliances. Furthermore, when inverter intended to supply inductive loads particularly electric motors and compressors whose startup current are much higher than the usual running current, then size of the inverter should be 3 times the capacity of those appliances to withstand surge current for short time. The ratings of inverter supplying isolated loads must be 25–30% bigger to handle the total amount of watts used at once.

2.2.6. *Battery sizing*

The batteries are important energy storage for grid and off-grid use. The type of battery recommended for solar PV system is deep-cycle battery. Deep-cycle batteries are designed to handle deep discharged and rapid recharged or cycle charged and discharged day after day for years. When sizing the batteries, important information needed is how much energy is consumed daily. When one is changing from grid power supply to renewable energy (solar) monthly, electric bill can be used to estimate daily energy consumption. Furthermore, in sizing the batteries, the intermitted energy supply from renewable energy has to be taken into consideration by estimating days of autonomy due to clouds or rain. The standard days of autonomy are estimated to be 3–5 days. To find out the size of the battery, calculate total watt-hours per day used by appliances and divide the total watt-hours per day used by 0.85 for battery loss. Then, divide the answer obtained by 0.6 for depth of discharge. Then, divide the answer obtained by the nominal battery voltage, and then, multiply the answer obtained with days of autonomy to get the required ampere-hour capacity of the deep-cycle battery.

2.3. Hydroenergy

2.3.1. Introduction

The geographical areas which are best for exploiting small-scale hydropower are those where there are steep rivers flowing all year round. In those areas, water turbines could be installed without a dam to generate electricity for home or community. A small or microhydroelectric power system can produce enough electricity for a village [19]. Small water turbine will produce power nonstop, as long as running water is available. Microwater current turbines are most suited for places where there is an almost a constant flow of water throughout the year. The underdeveloped and developing countries can use these techniques to provide electricity to remote places where transmission line cannot be connected easily or the cost becomes very high. Tanzania has extensive undeveloped hydroelectric resources mainly located in the southern region [20]. Religious centers (missionaries) and individuals in Tanzania are the first investors of microhydropower generation during the colonial period for mainly supplying power to a specific community or for their own use [21]. Geographically, the hydropower potentials of Tanzania are located in the rift valley escarpments which occur in the west, southwest, and northeast Tanzania. Studies show that 12 out 25 administrative regions of mainland Tanzania are blessed with min-hydropower resources but only three regions (Mbeya, Iringa, and Kilimanjaro) have at least managed to develop them.

Figure 11 shows the location of small hydropower sites which have been identified (green dot on the map) with different plant capacities up to 1 MW. Thus, the government is encouraging private investment in energy generation projects.

Majority of people in rural Tanzania are poverty prone and cannot afford the initial connection costs and the monthly bills. Rural electrification projects through grid extension and grid densification are associated with long transmission and distribution distances because of the sparse population as well as low load centers. In these market conditions, projects need government, multinational development agencies, NGOs, and the private sector to work together in order to design and create opportunities that respond to the needs of the local community.

In order to address these challenges, the government has established the Rural Energy Agency (REA) and the Rural Energy Fund (REF). The ongoing reforms in the power sector (liberalization and privatization) are anticipated to increase the interest of private firms investing in the hydropower generation.

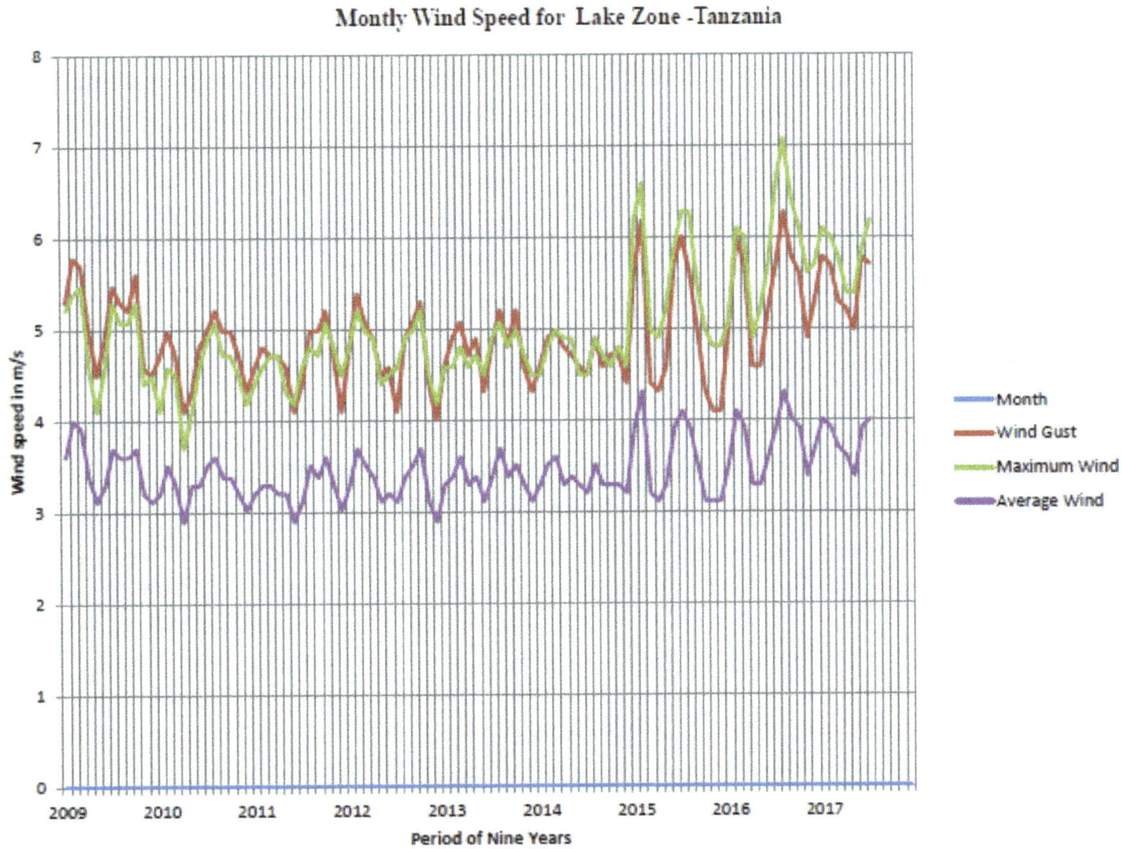

Figure 11. Hydropower potential in Tanzania.

2.3.2. *Functionality*

Most of the river water starts flowing from higher altitude to low altitude where there are lakes, ponds, or ocean. The river passes in the varying land pattern like steep slope, moderate slope, and nearly flat slope. Current turbines or hydrokinetic turbines are normally installed at the foot of steep slope, so that kinetic energy in water can be converted to mechanical energy for producing electricity. Depending on the nature of the river, that is, when the water flow is seasonal, then, water storage pond is constructed along the river, and when the flow is throughout the year, power-generating systems are directly installed along the flowing water in a river to enable turbine to rotate for producing electricity. Power produced by the moving water depends on water density, water head, and water discharge [22, 23].

2.3.3. *Cross-sectional area of natural water steam*

$$A_r = \left(\frac{a+b}{2}\right) \times \left(\frac{h_1 + h_2 + h_3 + \ldots h_k}{k}\right)$$

where a = width of top river in meter, b = width of bottom river in meter, and h = height in meter.

2.3.4. *The surface velocity*

A floating object, which is largely submerged, is located at the center of the stream flow. The time t (seconds) elapsed to traverse a certain length L (m) is recorded. The surface speed (m/s) would be the quotient of the length L and the time t.

$$V_{rs} = \frac{L}{t}$$

2.3.5. *The average flow speed*

To estimate the mean velocity, the above value must be multiplied by a correction factor that may vary between 0.60 and 0.85 depending on the watercourse depth and their bottom and riverbank roughness (0.75 is a well-accepted value).

$$V_r = 0.75 \times V_{rs}$$

The flow rate (Q):

$$Q = A_r \times V_r$$

2.3.6. *Internal diameter of penstock (Dp)*

$$D_p = 2.69 \times \left(n_p{}^2 \times Q^2 \times \frac{L_p}{H_g}\right)^{0.1875}$$

where n_p = manning coefficient, L_p = penstock length in meter, and H_g = gross head in meter.

2.3.7. *Penstock dimension*

Penstocks can be installed under and over the ground, depending on the nature of the ground. The penstock is built in nearly straight lines, with concrete anchor blocks at each bend and with an expansion joint between each set of anchors. The anchor blocks have been provided to resist the thrust and frictional forces caused by the penstock expansion and contraction. The straight section of the penstock lies in steel saddles, made from steel plates to reduce friction forces. For the part of penstock on the ground, it has been provided with supports of different heights depending on the nature of the ground. Spiraled metal sheet-welded steel pipes with flanges on both side have been considered, due to its reasonability in price and availability in different required sizes.

Recommended minimum wall thickness of penstock:

$$t_p = \left(\frac{D_p + 508}{400}\right) + 1.2$$

The vertical weight of penstock and water subjected to support:

$$F = (W_p + W_w) \times L_{ms} \times \cos\theta$$

where W_p = weight of penstock per meter in KN/m, W_w = weight of water per meter KN/m, L_{ms} = length of penstock between midpoints of each span, and θ = angle of pipe with horizontal.

Maximum length between the supports:

$$L_{mms} = 182.61 \times \frac{\left((D_p + 0.0147)^4 - D_p{}^4\right)^{\frac{1}{3}}}{P_w}$$

where P_w is weight of pipe full of water.

2.3.8. Powerhouse

The planned dimension of powerhouse is sufficient for safe operation and maintenance of all equipment included within it. A foundation is required on an adequate bearing soil stratum, with a concrete slab cast to provide a rigid base for the turbine and generator. A channel at the base slab is needed for outflow of water from the system. The powerhouse should be secured to prevent unauthorized access.

2.3.9. Power generated in watts

$$P_t = \rho g h_n Q \eta_t$$

where ρ = water density (1000 kg/m^3), h_n = head, and η_t = turbine efficiency.

2.3.10. Dump load

The load connected to hydropower varies time to time in a day that is peak hours and off-peak. During a day, peak hour's power produced is utilized by the connected load; therefore, the machine frequency would remain within the required limit. Furthermore, during the off-peak, load power consumed by the load is low compared to the power produced by the hydroturbine causing the machine to accelerate, hence the frequency increases. Therefore, to overcome this situation, a dump load in an electrical resistance heater (air or water) is installed to the system to dissipate excess power, so that the machine frequency is kept within the required limit. The size of dump load is usually equal to maximum power generated by the hydroturbine so that it can handle the full generating capacity of the microhydroturbine. Dump loads should be activated by the controller whenever power produced is not consumed by the load or grid (integrated system), to prevent machine accelerations which might result in system damage. Excess energy in the system is diverted to the dump load; it is envisaged most scheme offers excess power at all times.

2.3.11. Metering

In power plant, the quantities such as current, voltage, and frequency need to be monitored by measuring and displayed by the meters (whether analogue or digital). The load connected to hydropower needs specific magnitude of voltage and current. Operating load with the low magnitude of voltage and frequency results in efficiency reductions, while operating with higher values results in insulation failure, hence equipment damages. Therefore, several parameters of microhydroelectric system's performance and status are monitored by recording how much electricity is producing or has produced, and how much electricity is being consumed.

3. Government initiative in promoting renewable energy through small power producers

The first power plant was operated by diesel fuel in 1933 followed by 5-MW hydropower plant in 1936. These two power plants were supplied to the big town, leaving the rural area and small town with no electricity. Power generation capacity has been increasing in slow pace until 1959 with installed capacity reaching 17.5 MW. Furthermore, the isolated power plants operated by diesel were constructed in big towns all over the country.

The country has invested by constructing 21-MW Hale hydropower station in 1962, integrated with the existing power plants in north part of the county. For more integration, transmission line was constructed to integrate isolated power supply in the east part of the country. In 1969, there is an 8-MW Nyumba ya Mungu hydropower station on the headwaters of the Pangani River. In 1968, two hydropower plants generating units (50 MW each) to supply 100-MW Kidatu power station were installed. The installed capacity of the hydroelectric power station at Kidatu was doubled and reached 200 MW in 1980 [24].

All this investment in power generation plants was also accompanied by the construction of transmission to integrate the existing power plant as well as decommissioning diesel power plants.

The country has invested in power generation and distribution using state own company. Now, the country has the vision of becoming a middle-income country by 2025, with the electricity consumption of 490 kWh/capital. To support that vision, the Tanzanian government envisages the increase of electricity generation from 1357.69 MW in 2015 to 4915 MW by 2020 and improving electricity connections to 60% of the population from 18% in 2015. On average, the manufacturing sector will grow by over 10% per annum with its share in total exports increasing from 24% in 2014–2015 to 30% in 2020.

In order to meet energy requirement as well as the country's vision of becoming the middle-income country, energy policy was reviewed in order to invite independent power producer as one of the stakeholders in the electricity industry. The small power producers (SPPs) are private companies that develop renewable energy generation projects on a small scale (less than 10 MW). They are licensed to sell electricity either to local communities or to the national grid under a power purchase agreement with Tanzania Electric Supply Company Limited (TANESCO), a government-owned company, or to both TANESCO and the local communities.

4. Strategies of retaining youth in the rural areas

Tanzania has the population of 51.6 million; about 70% of this population lives in the rural area performing agriculture and animal keeping activities. They grow seasonal crops, food crops, fruits as well as cash crops and long-term crops. Most of the seasonal crops are planted during the raining season and harvested in dry season, while few of them practice irrigation agriculture. For the year with low rainfall, farmers get less harvest, hence need to buy food for their family; when rainfall is normal, they get more harvest. Moreover, they grow nonseasonal crops like cassava, cashew nut, banana as well as fruits like oranges, mangoes, and others. During harvesting period, the surplus food crops, fruits, and cash crops would be sold as raw due to the lack of storage facilities and value addition machinery because of lack of electricity. For example, coastal part of Tanzania is famous for the production of oranges. The situation in several markets during the orange harvesting is as seen in **Figure 12**.

Therefore, this lack of electricity in rural areas has caused poverty and also the quality of education is declining time to time as competent educators shifted to urban areas where there are good infrastructures for social life. This fact has discouraged young generation to stay in rural and they tend to migrate to urban leaving their community with no human resource to perform economic activities (farming and livestock keeping).

Therefore, women and youth in rural areas have to be empowered through effective participation in the management of their own social, economic, and environmental objectives by

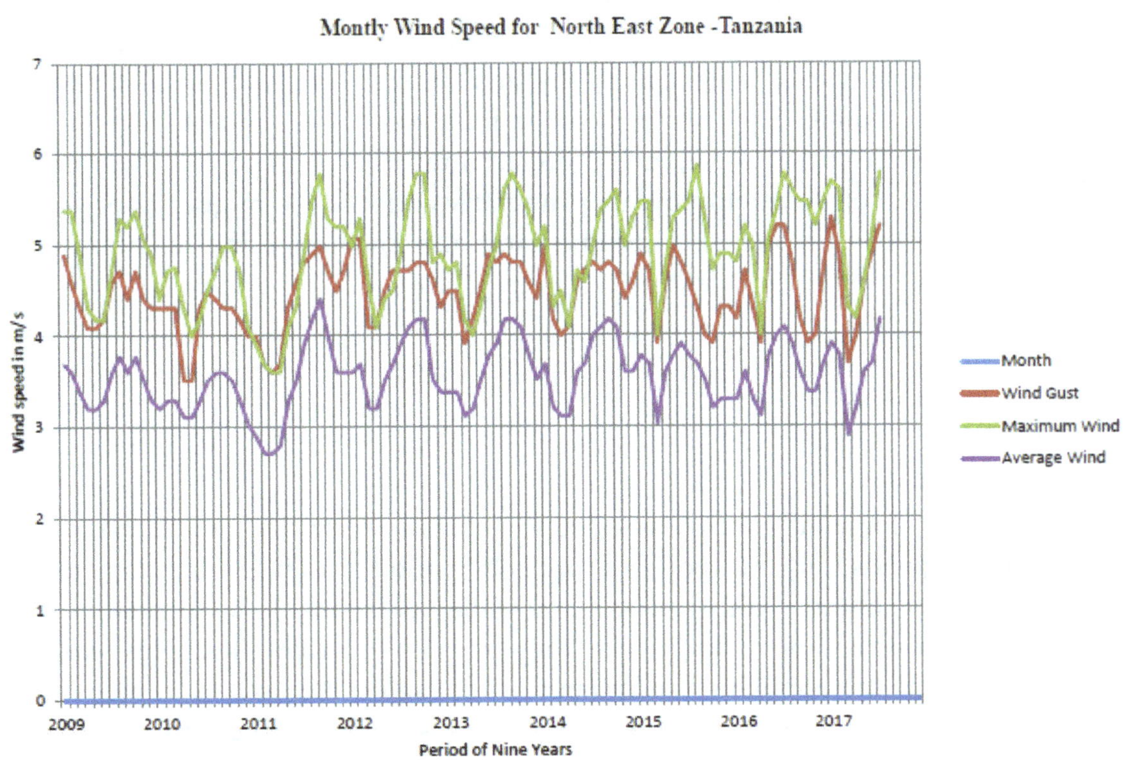

Figure 12. Oranges in the market.

establishing their own organizations such as local cooperatives and by applying the bottom-up approach.

Telecommunication in most of the rural areas is affected by the lack of reliable electricity to power microwave telecommunication links. The microwave links need reliable electricity in order to receive and transmit the information; therefore, due to the lack of electricity, the telecommunication providers failed to extend their service to rural areas due to economic viability. The few microwave links found in the rural area are powered by diesel generators which are very expensive.

On the other hand, lack of electricity causes the mobile subscribers to switch off their mobile phone to extend the cycle of recharging their mobile phone. Furthermore, the network is available in specific areas (especially higher elevations), where everybody is going there for communications. Therefore, an electrifying rural area will motivate telecommunication operators to install more microwave towers that will make the availability of network in the rural area as well as strengthen business among the communities in rural and urban areas.

Electrification of rural areas has to be done to stimulate income generation activities and value addition as well. Having the electricity in the rural area will keep busy the rural community by engaging themselves in several income generation activities in their area.

Electrification in the rural area will motivate establishment of small-scale industries for value addition to farm and animal's product. The value addition to milk product industry will motivate rural people to keep more cattle in the modern ways of getting more milk and manure. They will sell their milk to the processing industry where it can be processed for the export to the urban area. The other product is the manure which can be obtained by fermenting cow dung. The gas obtained can be used for heating purposes and the by-product is used as the manure in farms. Furthermore, the small industries like those making tomatoes as source, fabric from tomatoes, processing leather, food processing, and others will stimulate more investment in those sectors. Also, integration of rural areas with neighboring urban areas for the creation of rural off-farm employment can narrow down the migration of youth from rural to urban as well as expand opportunities and also encourage the retention of skilled people, including youth, in rural areas.

5. Conclusion

Renewable energy in Tanzania is not developed to generate electricity; hence, the community in the rural areas depends fully on wood fuel for heating and kerosene for lighting. This lack of electrical energy does not favor young people and skilled personnel to be retained in the rural areas for income generation activities; hence, they migrate to urban area. This situation has resulted in increase in the population in urban area engaging themselves in business, while production of food in rural areas declines due to the shortage of human resource, hence inadequate food in the country. The analyses of three major types of renewable energy that are wind, solar, and hydropower have shown great potential all over the country.

The information collected from Tanzania Metrological Agency shows that in several parts of the country, there is an adequate wind speed to be used for electricity generation, for grid integration, and stand-alone systems. For electricity generation, wind speed needed is 3 m/s and above which has been noticed. Thus, when embarking to the project, not only the mean wind speed but also the wind speed frequency distribution, commonly described by a Weibull distribution, has to be taken into account in order to estimate accurately the amount of electricity to be generated. The location with poor wind speed has blessed with other types of renewable energy; therefore, no location with no source of renewable energy.

Information collected for solar energy shows that there are good solar insolation all over the country of 4.0 and above. The country has two seasons that are rainy season and dry season. Normally, rainy season is between December and May, sometimes several hours of clouds in a day which does not affect solar system. In dry season, June–November, there is enough Sun for charging the systems. The solar insolation in the country is sufficient for grid integration as well as isolated and the roof-mounted solar system.

The information for hydropower in the country is located at the steep rift valley in western zone, northern zone, and west zone, which need to be developed. Most of the hydropotentials are pico, micro, and mini which can work at isolated systems and integrated system when there is grid networking nearby. Majority of people in rural Tanzania is poverty prone and cannot afford the initial connection costs and the monthly bills. Rural electrification projects through grid extension and grid densification are associated with long transmission and distribution distances because of the sparse population as well as low load centers. In these market conditions, projects need government, multinational development agencies, NGOs, and the private sector to work together in order to design and create opportunities that respond to the needs of the local community.

The government has put in place the policy to motivate small power producers (SPP), so that renewable energy power generation would be developed. The SPP would be licensed to sell electricity either to local communities or to the national grid under a power purchase agreement with Tanzania Electric Supply Company limited (TANESCO), a government-owned company, or to both TANESCO and the local communities.

Also, rural livelihoods have to be enhanced through effective participation of rural people and rural communities in the management of their own social, economic, and environmental objectives by empowering people in rural areas, particularly women and youth through organizations such as local cooperatives and by applying the bottom-up approach.

Author details

Urbanus F Melkior*, Josef Tlustý and Zdeněk Müller

*Address all correspondence to: melkiurb@fel.cvut.cz

Department of Electrical Power Engineering, Czech Technical University, Prague, Czech Republic

References

[1] https://www.usea.org/sites/default/files/event-/Tanzania%20Power%20Sector.pdf

[2] Kimambo C. Development of integrated water pumping and electricity generating wind system for remote applications. University of Dar Es Salaam; 2007

[3] URT. Sustainable Energy for all Rapid Assessment and Gap Analysis. 2013. http://www.se4all.org/sites/default/files/Tanzania_RAGA_EN_Released.pdf

[4] Mashauri A. A review on the renewable energy resources for rural application in Tanzania. Dar es Salaam, Tanzania: Electrical Engineering Department, Dar es Salaam Institute of Technology; 2011

[5] Melkior UF. Study of electrical power generation from windmill coupled with water pump. University of Dar Es Salaam; 2010

[6] World Bank Wind Resource Mapping in Tanzania Site Identification Report. 2015. http://pubdocs.worldbank.org/en/289831465287638659/Tanzania-Wind-Mapping-Site-Identification-Report-WB-ESMAP-July2015.pdf

[7] http://www.iea.org/publications/freepublications/publication/Wind_2013_Roadmap.pdf

[8] Al-shemmer T. Wind Turbine, 1st ed. 2010; ISBN: 978-87-7681-692-6

[9] Teubner Stuttgart BG. Grid Integration of Wind Energy Conversion Systems. 1996; ISBN: 0-471-97143

[10] Şen Z. Wind Velocity Vertical Extrapolation by Extended Power Law. https://www.hindawi.com/journals/amete/2012/178623

[11] Jäger K. Solar energy fundamentals, technology and systems. Delft University of Technology; 2014. https://courses.edx.org/c4x/DelftX/ET.3034TU/asset/solar_energy_v1.1.pdf

[12] United Republic of Tanzania. Renewable Energy Policies and Practice in Tanzania

[13] http://www.nclark.net/photosynthesis.pdf

[14] Nzali AH. Insolation energy data for Tanzania. In: International Conference on Electrical Engineering and Technology; the University of Dar es Salaam; 2001. pp. EP26-EP32

[15] Al-Tameemi MA, Chukin VV. Global water cycle and solar activity variations. Journal of Atmospheric and Solar - Terrestrial Physics. 2016;**142**:55-59

[16] Hart Hubris, the troubling science, economics, and politics of climate change, Compleat Desktops Publisher (2015). ISBN: 9780994903808

[17] Luqman M, Ahmad SR, Khan S, Ahmad U, Raza A, Akmal F. Estimation of solar energy potential from rooftop of Punjab government servants cooperative housing society Lahore using GIS [article ID:56795]

[18] https://gcep.stanford.edu/pdfs/assessments/solar_assessment.pdf

[19] Zainuddin H, Yahaya MS, Lazi JM, Basar MFM, Ibrahim Z. Design and development of pico-hydro generation system for energy storage using consuming water distributed to houses. International Journal of Electrical and Computer Engineering World Academy of Science, Engineering and Technology. 2009-11-23;**3**:150-155

[20] https://www.esmap.org/sites/esmap.org/files/01-KEF2013-REM_Gratwicke_Rift%20Vallery%20Energy.pdf

[21] https://www.esmap.org/sites/esmap.org/files/TEDAP%20SPPs%2011-18.pdf

[22] Kumar A, Schei T, Ahenkorah A, Caceres Rodriguez R, Devernay JM, Freitas M, Hall D, Killingtveit A, Liu Z. Hydropower. In: IPCC Special Report on Renewable Energy Sources and Climate Change Mitigation. Cambridge, United Kingdom and New York, NY, USA: Cambridge University Press; 2011

[23] Nasir BA. Design of micro-hydro electric power station. International Journal of Engineering and Advanced Technology (IJEAT) ISSN: 2249-8958. June 2013;**2**(5)

[24] http://www.ewura.go.tz/wp-content/uploads/2017/01/Power-System-Master-Plan-Dec.-2016.pdf

10

Reliability and Aging Analysis on SRAMs Within Microprocessor Systems

Taizhi Liu, Chang-Chih Chen and Linda Milor

Abstract

The majority of transistors in a modern microprocessor are used to implement static random access memories (SRAM). Therefore, it is important to analyze the reliability of SRAM blocks. During the SRAM design, it is important to build in design margins to achieve an adequate lifetime. The two main wearout mechanisms that increase a transistor's threshold voltage are bias temperature instability (BTI) and hot carrier injections (HCI). BTI and HCI can degrade transistors' driving strength and further weaken circuit performance. In a microprocessor, first-level (L1) caches are frequently accessed, which make it especially vulnerable to BTI and HCI. In this chapter, the cache lifetimes due to BTI and HCI are studied for different cache configurations, namely, cache size, associativity, cache line size, and replacement algorithm. To give a case study, the failure probability (reliability) and the hit rate (performance) of the L1 cache in a LEON3 microprocessor are analyzed, while the microprocessor is running a set of benchmarks. Essential insights can be provided from our results to give better performance-reliability tradeoffs for cache designers.

Keywords: reliability analysis, SRAM stability, cache configurations, microprocessors, semiconductor microelectronics, very-large-scale integration (VLSI)

1. Introduction

As smaller technology nodes bring significant benefits like more density and lower power consumptions, they also pose significant reliability challenges. Not only do the manufacturing variations make the resulting transistors unreliable at low-voltage operation but also they take less time to wear out, making them more prone to failures in the field. The increasing reliability concerns hold for all types of microelectronic devices from electronics used in aerospace applications where reliability requirement is extremely critical, to mobile devices where product reliability can strongly affect market share.

BTI and HCI are two of the most dominating wearout mechanisms that increase the threshold voltage (V_{th}) of a transistor. As a result of BTI and HCI, the driving strengths of the aged transistors are weakened, which eventually could cause timing violations and faulty operation. During the static-stress window when a transistor is kept ON, BTI kicks in. There are two forms of BTI: Negative BTI (NBTI) and Positive BTI (PBTI). NBTI affects the threshold voltage of a PMOS transistor when its gate is applied LOW; and PBTI affects the threshold voltage of a NMOS transistor when its gate is applied HIGH. On the other hand, HCI happens when a transistor flips from being OFF to ON or vice versa. Therefore, HCI is more acute to those transistors that switch frequently.

In a modern microprocessor, static random access memories (SRAM) take the majority of the transistors, and thus the reliability of the SRAM cells is essential for circuit designers. Moreover, the first-level (L1) data cache is frequently accessed (read and written), making it very vulnerable to HCI. But at the same time, it also stores data for a significant amount of time, making it also vulnerable to BTI. Besides, cache efficiency is one of the most important characteristics for microprocessor system performance. Basically, for microprocessor designers, it is very important to understand both the performance and the reliability of the cache systems. There are many prior works [1–3] focused on cache architecture to improve cache efficiency. However, when different advanced techniques are used to achieve higher performance, it is still unknown how the reliability of the cache is changed. In this chapter, the failure probability of the L1 data cache is investigated for a LEON3 microprocessor when different design configurations are applied: associativity, cache line size, cache size, and replacement algorithm. We analyzed the impact of cache configurations on failure rates and cache efficiency so that cache designers can achieve performance-reliability tradeoff according to their design budgets (area, power, lifetime, etc.). We also study the impact of error correcting codes (ECC) on cache reliability.

BTI and HCI cause driving-strength mismatch in a traditional six-transistor (6T) SRAM cell. Because SRAM stability is extremely sensitive to transistor mismatches, BTI and HCI pose a significant problem to SRAM reliability [4–6]. In [7–9], the authors analyzed SRAM stability by assuming two ideal stress conditions, that is, static stress (0% or 100% duty cycle) and alternating stress (50% duty cycle). However, the realistic stress conditions of the SRAM cells really depend on customer usages (workload). In [10–13], the authors estimated the SRAM degradation due to BTI based on the realistic stress conditions considering the actual workload. On the other hand, the impact of the HCI effect on SRAM stability is not as studied as BTI because BTI is usually dominant due to its frequency independence. However, HCI is becoming more concerning as operating frequencies of nowadays chips are GHz-level [14, 15]. Some prior arts have investigated the HCI effect on SRAM cell stability [16, 17], and in [16], the simulation results are even compared with silicon experimental results.

Other research efforts have focused on balancing the amount of time that logic '0' and '1' values are stored in the cells with the aim to provide a BTI-optimal duty cycle distribution [18, 19], and by implementing redundancy into the cache design to combat BTI-induced wearout [20]. Gunadi et al. [19] also proposed to mitigate the HCI degradation by providing a uniform distribution of cache accesses across sets.

In this chapter, we stress SRAM cells under different stress conditions and analyze the SRAM stability due to BTI and HCI. As a case study, the L-1 data cache of a state-of-art microprocessor (LEON3) is studied, and cache reliability and cache efficiency are analyzed by considering the realistic workload when the microprocessor is running a set of benchmarks.

2. Device-level wear-out mechanisms

We first model BTI and HCI at the device level and then abstract the models to the system level.

2.1. NBTI/PBTI

Negative BTI, as known as NBTI, is the degradation for PMOS transistors when negative gate-to-source voltage is applied. Positive BTI, known as PBTI, is the degradation of NMOS devices under positive gate-to-source voltage. Both NBTI and PBTI can cause an increase in the threshold voltage and the consequent decrease in drain current and transconductance of a MOSFET.

According to trapping/de-trapping theory [21], the threshold voltage shift (ΔV_{th}) due to BTI is modeled as a function of time under DC stress (t_{DC}):

$$\Delta V_{th}(DC) = \phi(T, E_F)(A + \mathrm{B}\ln(t_{DC})) \tag{1}$$

where ϕ is proportional to the number of available traps and is a function of temperature, T, and the Fermi level, E_F, and A and B are constants. The temperature dependence of BTI is incorporated in ϕ with the Arrhenius relationship:

$$\phi(T, E_F) = \phi_0 g(E_F) e^{-E_a/kT} \tag{2}$$

where k is a constant, T is temperature, and E_a is the activation energy. Since the frequency dependency of BTI has been considered as relatively insignificant, especially for low-frequency signals [22], it is not included in this work. However, the duty cycle, α, can affect the ΔV_{th}, and it is incorporated as an effective Fermi level, where $E_{F,eff} = \alpha E_{F,on} + (1 - \alpha) E_{F,off}$. Here, $E_{F,on}$ and $E_{F,off}$ are the Fermi levels when the transistor is ON and OFF, respectively. The duty cycle accounts for the time under stress, t_{stress}, and the recovery time, t_{rec}, since $\alpha = t_{stress}/(t_{stress} + t_{rec})$. The function $g(\alpha)$ in Eq. (2) is a nonlinear function of α, which has $g(1) = 1$ and $g(0) = 0$ [21]. Overall,

$$\Delta V_{th} = \phi_0 e^{-E_a/kT} g(t_{stress}/(t_{stress} + t_{rec})) \cdot (A + Bln(t_{stress} + t_{rec})) \tag{3}$$

where ϕ_0 is a constant. The constants were obtained from the experimental results in [23].

2.2. HCI

Hot carrier injection (HCI) is the phenomenon where electron or a "hole" gains sufficient kinetic energy to overcome a potential barrier necessary to break an interface state to be injected into the gate oxide. HCI is one of the mechanisms that adversely affect the reliability of semiconductors of solid-state devices. More specifically, some of the device parameters such as the threshold voltage, channel mobility, drain saturation current, and transconductance can be degraded due to HCI. HCI was a major concern for NMOS transistors historically, and the HCI effect on PMOS transistors was relatively negligible. This was because holes have a smaller impact ionization rate than electrons, and the $Si - SiO_2$ barrier for holes is also higher than electrons. However, researchers have recently observed HCI effects on PMOS transistors [24].

As hot carriers are generated during switching of the transistors, the HCI effect is directly proportional to the switching frequency. In this chapter, we used the predictive HCI lifetime models for long-term performance-degradation simulations, where the ΔV_{th} degradations due to HCI during stress time are modeled as [25–27]:

$$\Delta V_{tp/tn} = A_{HCI} (r_{trans} t_{stress} t_{trans})^n \tag{4}$$

where t_{stress} is the stress time, r_{trans} is the frequency-dependent transition rate, t_{trans} is the transition time, and A_{HCI} is a constant that depends on the inversion charge, the trap generation energy, the hot electron mean free path, and other process-dependent factors [28, 29].

3. SRAM stability

3.1. SRAM cell

Each SRAM cell can store one bit, and it is usually implemented using six transistors, which is well known as 6T SRAM cell. The structure of a 6T SRAM cell is shown in **Figure 1**. The core of the cell is formed by two CMOS inverters (the four labeled transistors in **Figure 1**), where the output potential of each inverter is fed as input into the other. The formed feedback loop stabilizes the inverters to their respective state. Besides the inverter loop, the remaining unlabeled two transistors in **Figure 1** are the access transistors, which are controlled by the word and bit lines, WL and BL, respectively. WL and BL are used to read and write from or to the cell. When the word line (WL) is low, the access transistors are turned OFF, and the cell is in standby mode. When reading, the word line (WL) is HIGH and the access transistors are

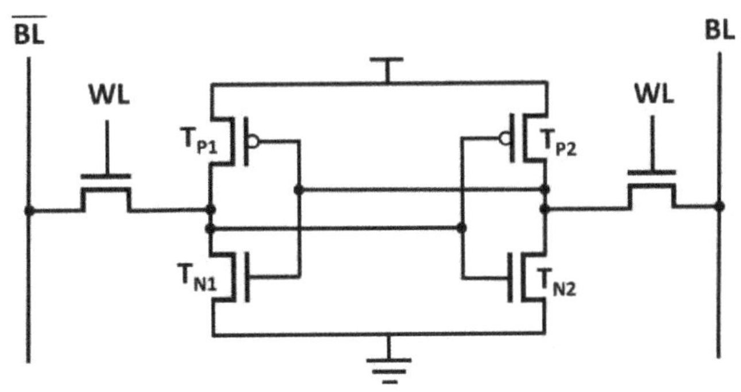

Figure 1. A typical 6T SRAM cell.

ON to allow the stored bit reflected at the bit lines. When writing, the word line (WL) is also HIGH to turn access transistors ON, and the asserted bit lines are strong enough to write the data into the inverter loop.

For the 6T SRAM cell mentioned above, all the transistors will be affected by the HCI effect during a write access when the stored bit changes. For the BTI effect, it happens when the stored bit is stable and the transistors are in static stress. More specifically, when the stored bit is a '0,' the PMOS transistor T_{P1} and the NMOS transistor T_{N2} are stressed because they are turned ON, meaning they are undergoing NBTI and PBTI, respectively. On the other hand, if a '1' is stored, the other two transistorsT_{P2} and T_{N1} are turned ON, and they are suffering from NBTI and PBTI, respectively. It is worth noting that, when one pair of transistors (T_{P1} and T_{N2}, for example) is under stress and undergoing BTI, the other pair (T_{P2} and T_{N1}) is not under stress and is under recovery from BTI degradation. However, overall, these transistors that form the inverter loop (T_{P1}, T_{N2}, T_{P2}, and T_{N1}) are continuously aging regardless of whether the cell is being read or write [30]. For the access transistors, they are only affected by BTI during the SRAM cell is being accessed (when WL is HIGH). Thus, the access transistors are much less sensitive to BTI than the inverter-loop transistors. In this chapter, we focus on the aging of the inverter-loop transistors.

3.2. Extraction of activity, temperature, IR-drop profiles

BTI and HCI effect not only depends on the time that the device is under stress but also depends on temperature. The time that the device is under stress is referred to as stress time in the following chapter. For BTI, the stress time is proportional to the duty cycle, that is, for a NMOS transistor, the stress time is equal to the total time (that the circuit is working) multiplied by the percentage so that the gate voltage is HIGH, while for PMOS transistors, it is equal to the total time multiplied by the percentage so that the gate voltage is LOW. For HCI, the stress time is proportional to the number of switching.

For the memory block within a microprocessor, it is not feasible to run SPICE simulations to get the activity (duty cycle, switching) profile of each SRAM cell. In our work, we utilize a FPGA emulation system to simulate the microprocessor. Being doing so, we are able to run benchmarks on the microprocessor and extract the activity profile in an efficient manner. Our framework to extract activity profiles is shown in **Figure 2**, which also includes the further steps to extract thermal profiles. To extract the activity profile, we synthesized the hardware RTL of the design into an FPGA and placed counters at the I/O ports of the data cache. The placed counters can track both the state probabilities (duty cycle) and the toggle rates at the I/O ports when the microprocessor is running benchmarks. The state probability is the probability of a net at each logic state, that is, logic '0' and logic '1,' and the toggle rates are the number of toggles that a net has during a unit period, for example, 1 ns. The extracted activities (state probabilities and toggle rates) were then used for activity propagation to get the complete activity profile of all the SRAM cells.

Besides activity extraction, the thermal profile throughout the microprocessor is also extracted. Moreover, because the SRAM stability strongly depends on voltage, we also consider the

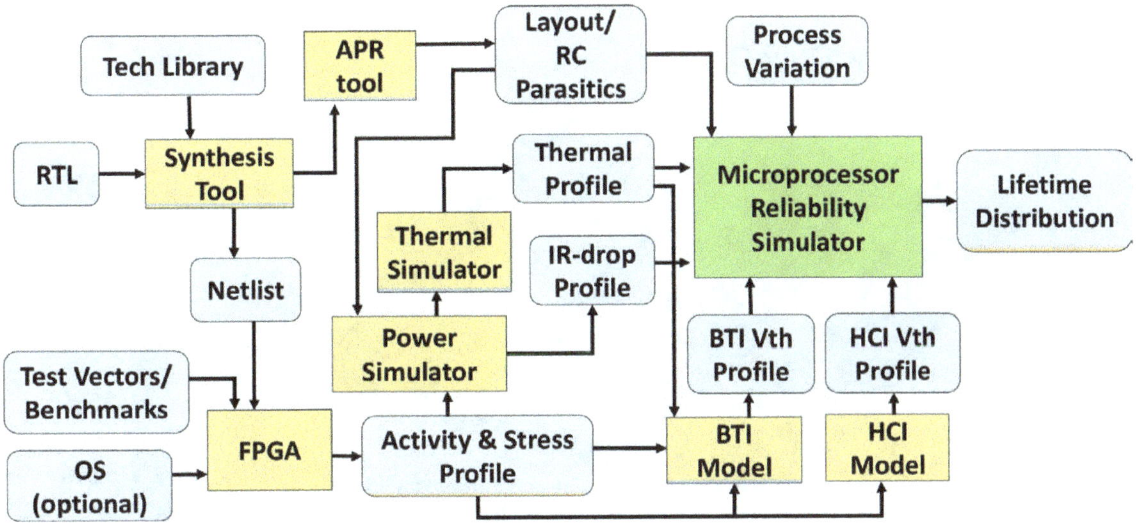

Figure 2. The FPGA-based aging assessment framework, which is used to extract the duty cycle/toggle-rate profiles, temperature profile, and the IR-drop profile.

impact of IR-drop in our work. As shown in **Figure 2**, the netlist was used for layout generation, and then RC parasitics from the layout, along with the activity profile, are fed to extract the power profile and the consequent thermal profile, using the power simulator [31] and the thermal simulator [32], respectively, for every module block of the microprocessor system.

In this chapter, we used the open-source microprocessor called LEON3 [33] as a case study. LEON3 is well known for space applications with high-level reliability requirement. We have implemented LEON3 with superscalar abilities on a commercial 90 nm technology process. The logic part of the LEON3 core includes a 32-bit multiplier (MUL), a 32-bit divider (DIV), a 32-bit general purpose integer unit (IU), and a memory management unit (MMU). The memory part of the LEON3 core consists of data caches (D-Caches) and instruction caches (I-Caches), cache tag units (Dtags and Itags), and window-based register file (RF). In this chapter, we focus our analysis on L1 D-Caches due to its importance to microprocessor performance and its high sensitivity to aging effects. The proposed method is applicable to other memory blocks as well.

Standard benchmarks from MiBench [34] were used as the microprocessor applications. **Figures 3** and **4** show the distributions of the state probabilities and the transition rates, respectively, of the data cache, when the microprocessor is running a standard benchmark. **Figure 5** shows the average temperature distribution and average IR-drop distribution when the microprocessor is running a standard benchmark.

3.3. SRAM stability degradation analysis under BTI and HCI

In this chapter, several performance metrics were used to characterize SRAM stability, including the read and retention static noise margins (SNMs), the read current (I_{READ}), the minimum retention voltage (Vdd-min-ret), and the write margin. SNM is a key figure of merit for an SRAM cell. It is the minimum DC noise voltage necessary to change the state of an SRAM cell and can be extracted by nesting the largest possible square in the two voltage transfer curves

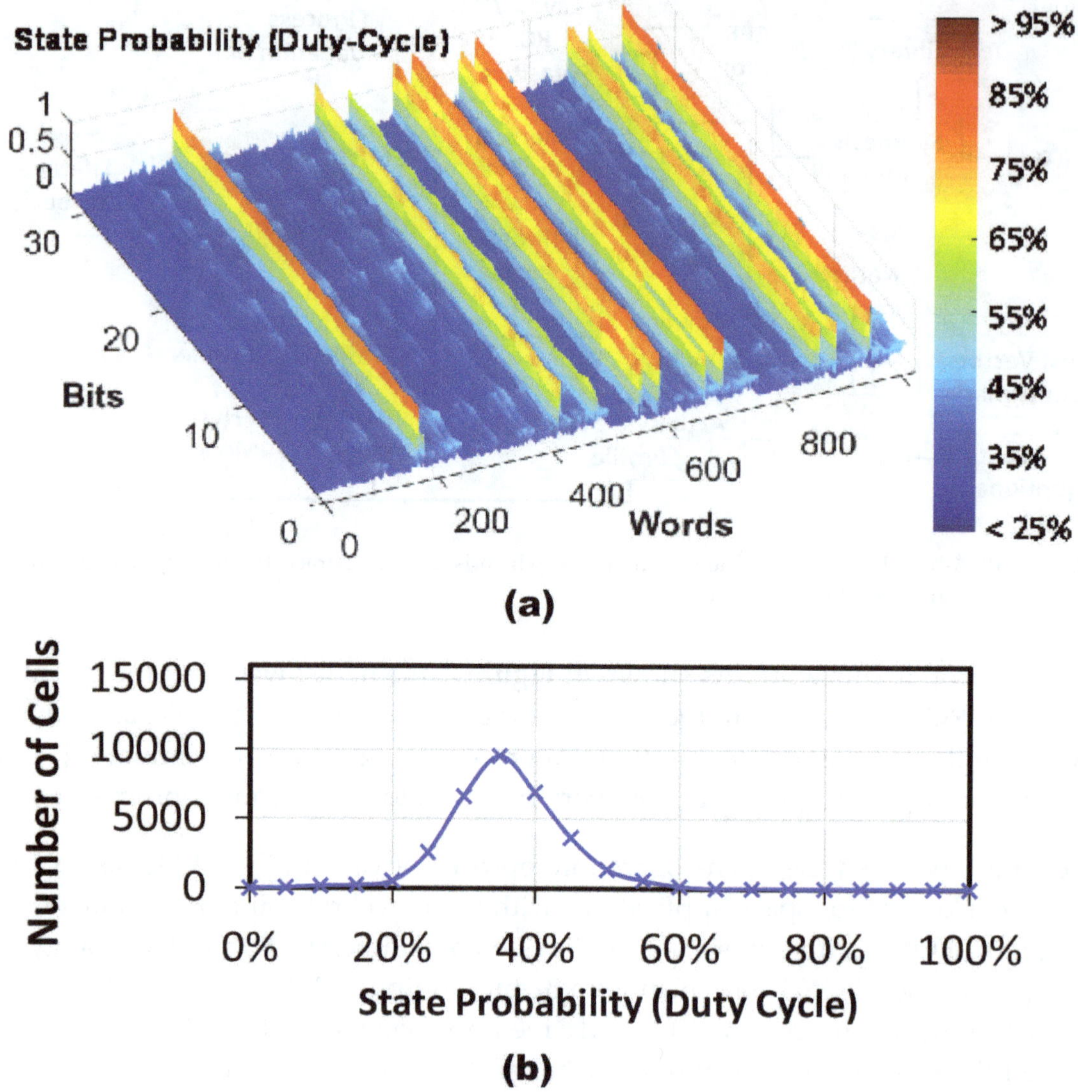

Figure 3. (a) The distribution of state probability for the 32 KB data cache shown in 1024 words and (b) the histogram of the state probability distribution in the number of SRAM cells.

(VTC) of the involved CMOS inverters [35]. The read SNM is measured when the access transistors are turned ON, while for the retention SNM, the access transistors are OFF. I_{READ} is the current flowing through pull-down transistors during a read access, and it is inversely proportional to access time. Vdd-min-ret is the minimum supply voltage that an SRAM can retain the stored bit. The write margin is the minimum voltage needed to flip the state of the cell, with the access transistors are ON. The lifetime calculations in this chapter are based on the following assumption: when any of these four metrics mentioned above has degraded to a predefined threshold, the SRAM cell is said to have failed and thus the lifetime of the cell is calculated.

In this chapter, the process variations of two important parameters, channel length and threshold voltage, are included, assuming they follow normal distribution with standard deviation equal to 10% of their corresponding nominal values.

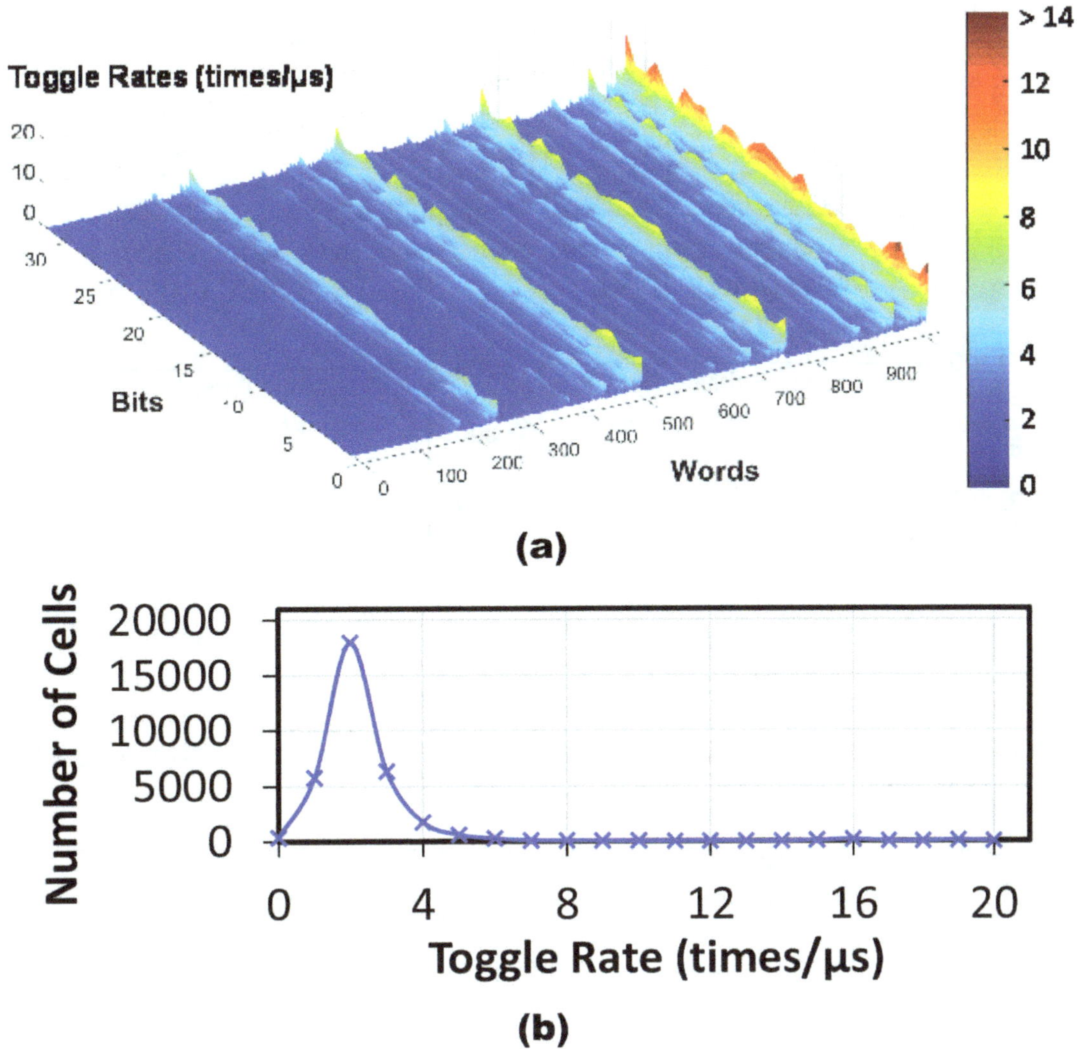

Figure 4. (a) The distribution of transition rate for the 32 KB data cache shown in 1024 words and (b) the histogram of the transition-rate distribution in the number of SRAM cells.

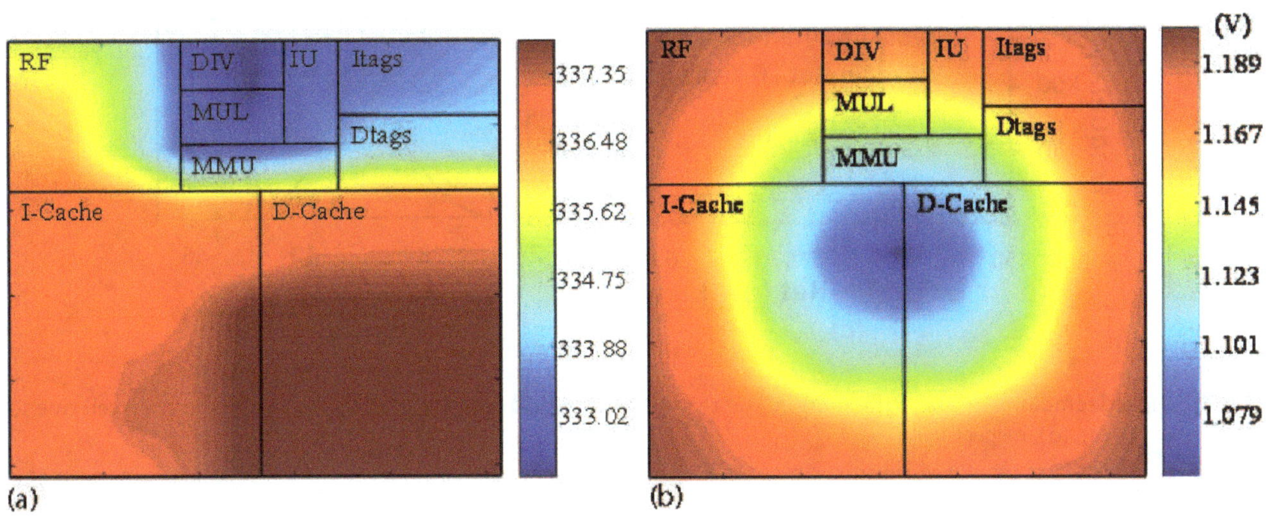

Figure 5. (a) The average temperature distribution and (b) the average IR-drop distribution of the microprocessor while running a standard benchmark.

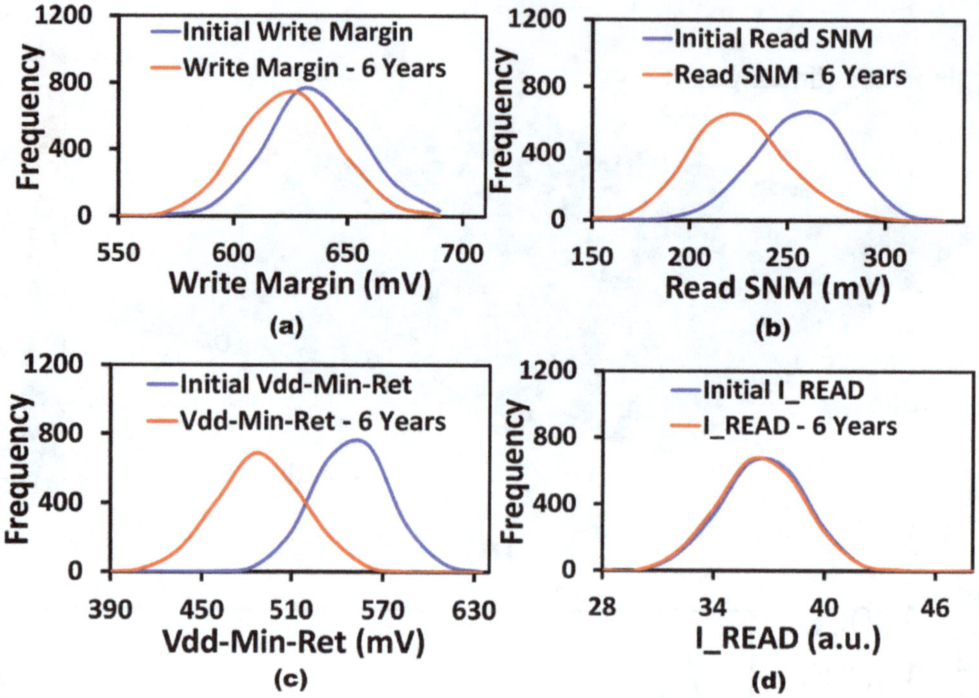

Figure 6. The degradation of the write margin, the read SNM, the Vdd-min-ret, and the I_{READ} of a memory cell due to BTI shown in (a)–(d), respectively.

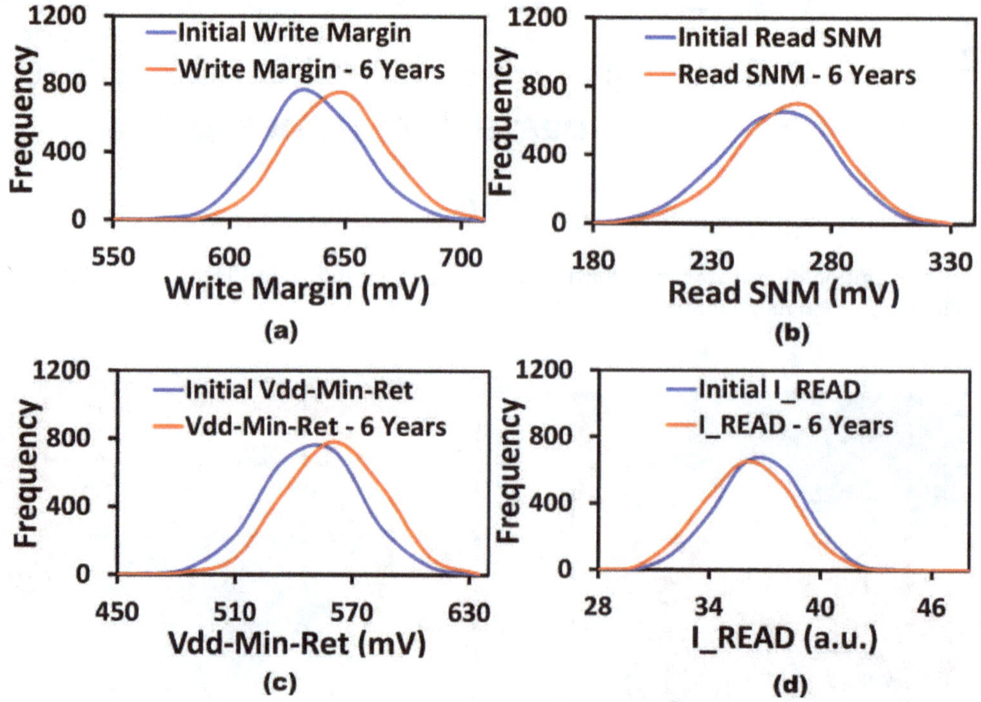

Figure 7. The degradation of the write margin, the read SNM, the Vdd-min-ret, and the I_{READ} of a memory cell due to HCI shown in (a)–(d), respectively.

Figures 6 and **7** show the degradation of the read SNM, the write margin, the Vdd-min-ret, and the I_{READ} of a memory cell due to BTI and HCI, respectively. As it is seen from **Figure 6**, BTI severely degrades the read SNM as well as the write margin. The Vdd-min-ret is also affected, while the I_{READ} is relatively unaffected. On the one hand, HCI, as shown in **Figure 7**, only degrades I_{READ} and improves the other three cell performances. This is because the cell becomes increasingly skewed under BTI as some devices degrade more than the others. This leads to impaired noise immunity. On the other hand, all the devices undergo the same stress due to HCI, as explained in Section 3.1.

4. Lifetime analysis

4.1. Memory cell lifetime characterization

To estimate the SRAM lifetimes due to BTI and HCI, the activity profile, thermal profile, and IR-drop profile of the memory were collected by the framework as shown in Section 3.2. The stress and thermal profiles are fed into the BTI and HCI models described in Section 2 to obtain the threshold voltage degradation. Then, the thermal profile, IR-drop profile, the BTI and HCI threshold voltage degradations, together with process parameter variations, were used to analyze the degradation of SRAM stability via Monte Carlo SPICE simulations (2000 samples for each Monte Carlo run). As mentioned in Section 3.3, an SRAM cell is assumed to have failed when any of the aforementioned four metrics degrades the predefined threshold levels. Then, the lifetime of the SRAM cell is obtained by interpolating the two time stamps where the failure happens in between. To characterize the cell lifetime, the cell is simulated 2000 times for each of the time stamps in SPICE. The time stamps basically define the level of BTI/HCI degradations, that is, the BTI/HCI-induced threshold voltage shifts are back annotated to the SPICE netlist for Monte Carlo simulations.

If we run Monte Carlo SPICE simulations for each cell for each time stamp, it would be very time-consuming and not practical. To address the large number of cells, the state probabilities and toggle rates are partitioned into 21 stress states (0%, 5%, 10%, …, 95%, 100%) for BTI and HCI, respectively. This strategy can dramatically reduce the cost of SPICE simulation time while not giving up too much accuracy. It is straightforward to assume that cells from the same stress state share the same state probability and the same toggle rate. Furthermore, all the cells in one stress state share the same lifetime distribution.

For BTI, the stress states are partitioned by state probabilities. The 0% stress state means that 0% of time the cell is storing a '1,' while the 100% stress state means a '1' is stored all the time. For HCI, the stress states are the percentage of the maximum toggle rate that we observed, that is, 0%, 5%, 10%, …, 100% of the maximum toggle rate. **Figures 3(b)** and **4(b)** show an example for the stress-state distribution for BTI and HCI, respectively, for a 32 KB data cache. The stress-state distribution not only depends on the benchmark that is running but also depends on the configuration of the cache system. We will discuss this impact in Section 5.

As process variations are considered, the lifetime of each SRAM cell is now a distribution rather than a value. With Monte Carlo simulations, the lifetime distribution is computed for each stress state. Importance sampling [36] was employed to have sufficient samples for the tail part of the distribution. **Figures 8** and **9** show the lifetime distributions for five representative stress

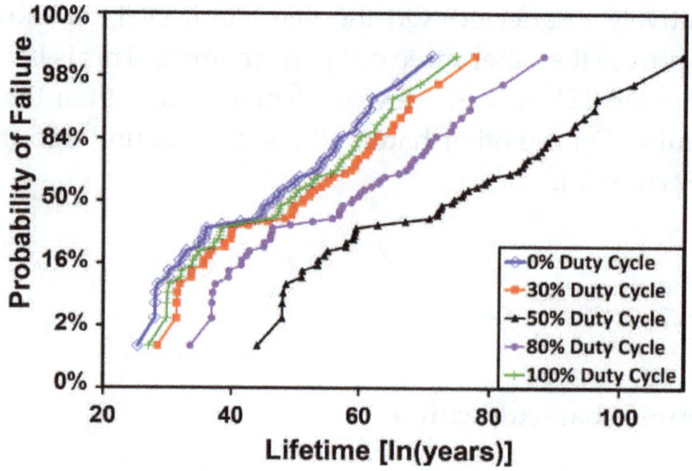

Figure 8. The BTI lifetime distribution of an SRAM cell when it is under a specific duty cycle stress state. Five duty cycle stress states are shown as follows: 0%, 30%, 50%, 80%, and 100%.

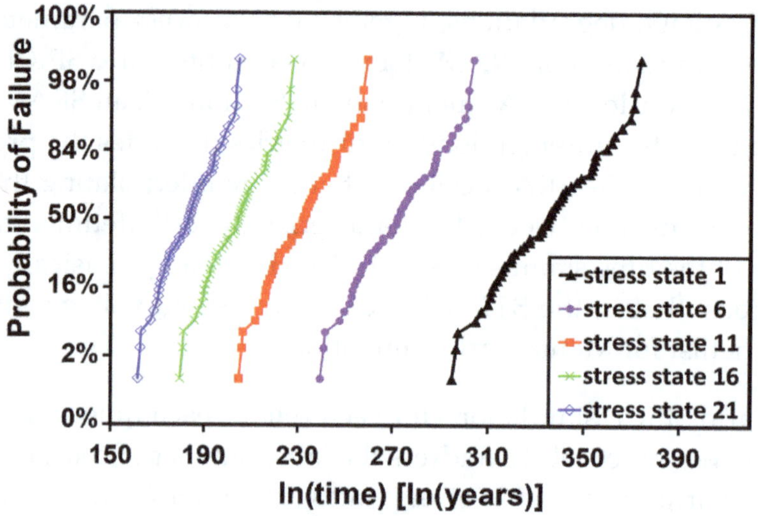

Figure 9. The HCI lifetime distribution of an SRAM cell when it is under a specific toggle-rate stress state. Five toggle-rate stress states are shown: State N means a toggle rate of N times/μs.

states, for BTI and HCI, respectively. As shown, for BTI, 50% stress state has the best lifetime, while for HCI, the lowest switching rate results in the best lifetime.

Log-normal distribution is the best fit for the lifetimes in **Figures 8** and **9**. Once the fitted log-normal distributions are determined, it is straightforward to obtain the failure rate of an SRAM cell, PF_{bit}, as a function of time, t:

$$PF_{bit} = Probability\ of\ (Lifetime < t). \tag{5}$$

Then, the failure probability of a word can be calculated, assuming no error correction codes:

$$PF_{word} = 1 - \prod_{i=1}^{N} \left(1 - PF_{bit_i}\right) \tag{6}$$

where N is the number of bits in one word, PF_{word} is the failure probability of a word, and PF_{bit} is the failure probability of a bit. Without ECC, we can safely assume that if there is one cell fails to work, the whole memory system will fail. It is then straightforward to get the failure probability of the whole SRAM block:

$$PF_{SRAM} = 1 - \prod_{i=1}^{N_{word}} \left(1 - FP_{word_i}\right) \quad (7)$$

where N_{word} is the number of words, PF_{SRAM} is the failure probability of of the memory block, FP_{word_i} is the probability of failure of i -th word. As PF_{bit} is a function of time, PF_{word} is also a function of time, and so is PF_{SRAM}.

The inclusion of error correcting codes can detect and correct the internal data corruption in SRAMs. In this chapter, BCH codes [37] were used, which consumes seven additional bits per word and can correct one bit per word. With ECC, for a word containing N bits (including ECC), the failure probability of a word, F_{word}, is different from Eq. (6):

$$PF_{word} = 1 - \prod_{i=1}^{N} \left(1 - PF_{bit_i}\right) - \sum_{j=1}^{N} \left[PF_{bit_j} * \prod_{i \neq j} \left(1 - PF_{bit_i}\right)\right] \quad (8)$$

In LEON3, the word size is $N = 32$ for the data cache without error correcting codes (ECC). With ECC, the word size is $N = 39$. Note that Eqs. (5) and (7) are the same for with ECC and without ECC.

5. Performance-reliability analysis for different cache configurations

In this section, we study the impact of cache configurations on cache reliability. Four categories are considered, including cache associativity, cache size, cache line size, and the replacement algorithm. The cache hit rates are also presented along with the cache reliability to analyze the performance-reliability tradeoffs. Besides, we also show the impact of error correction codes (ECC) on cache reliability.

Six benchmarks from MiBench [34] are tested: Qsort, SHA, CRC32, FFT, Basicmath, and Dijkstra. Qsort benchmark implements the classical Qsort algorithm on a large array of strings. SHA benchmark produces a 160-bit digest for a given input by using the classical secure hash algorithm. CRC32 benchmark performs a 32-bit Cyclic Redundancy Check (CRC) to detect errors in data transmission. FFT benchmark performs a fast Fourier transform on an array of data. Basicmath benchmark has many basic mathematical calculations, which usually do not have dedicated hardware support in embedded processors. Finally, the Dijkstra benchmark implements the well-known Dijkstra's algorithm to get the shortest path between every pair of nodes on a large graph, which is stored in an adjacency matrix.

The state-probability (duty cycle) distributions are shown in **Figure 10**, for each of the six benchmarks mentioned above. It can be obviously seen that the distributions are leaning to the left. It is because in data cache memory, logic '0' is more dominant than logic '1' [38]. In fact, memory is typically initialized to all '0's when allocated. This means, even if the benchmark is

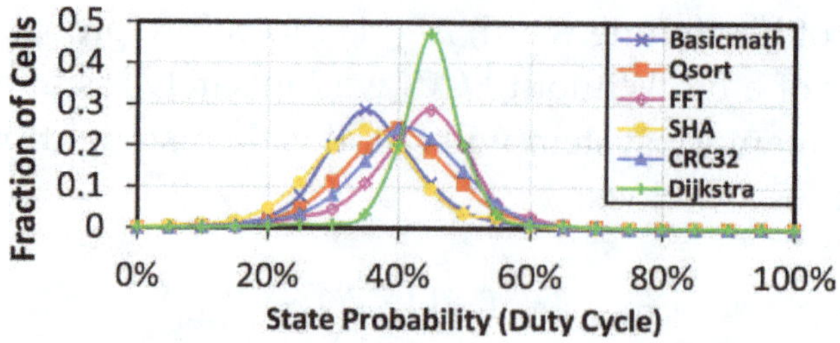

Figure 10. The duty cycle distributions of SRAM cells in a two-way 32 KB data cache while the microprocessor is running six different benchmarks.

writing a '0' and '1' to any bit with equal likelihood, logic '0' is always stored longer than logic '1'. There are some other reasons for '0' being stored longer, including false Boolean values and NULL pointers are represented with '0's, and most data in dense-form sparse matrices are '0's [39].

In our setup, the microprocessor is running at 250 MHz frequency. For this level of frequency, BTI is dominant and the HCI effect has a smaller impact. This is because that BTI is independent of frequency, while HCI is frequency dependent and 250 MHz is not a very high frequency. However, the HCI effect would be more impactful if the microprocessor is working at higher frequencies.

The overall failure probability of the SRAM block is calculated based on the following equation:

$$PF_{SRAM.Overall} = 1-(1-PF_{SRAM.BTI}) * (1-PF_{SRAM.HCI}). \tag{9}$$

where $PF_{SRAM.BTI}$ is the failure probability due to BTI, and $PF_{SRAM.HCI}$ is the failure probability due to HCI.

5.1. Associativity

There are three types of cache associativity: fully associative, direct mapped, and n-way set associative. For fully associative, data could be anywhere in the cache, making it very expensive to implement as it must check the tag of every cache line. For direct mapped, data can only go to a single cache line in the cache based on the memory address of the data. Set associative cache is a trade-off between direct mapped cache and fully associative cache. The cache is divided into 'n' sets, and each set contains a number of cache lines. Four-way set associative means the cache is divided into sets that can fit four blocks each, while a two-way set associative means each set can hold two blocks. From this perspective, a fully associative cache of m cache lines is m-way set associative, and a direct mapped cache is actually 1-way set associative. Although higher associativity can achieve higher hit rate, it is more expensive in terms of timing and area cost.

In our work, we have implemented the LEON3 data cache with three different associativities: 1-way, 2-way, and 4-way. Other configurations are kept the same: 16-byte cache line size, 32 KB cache size, and LRU replacement algorithm.

Figure 11 shows the failure probability of the whole data cache for two illustrative benchmarks: Basicmath and Dijkstra (other benchmarks have a similar trend). The hit rates for

1-way, 2-way, and 4-way associativity for Basicmath are 96.12%, 96.33%, 96.36%, respectively. For Dijkstra, they are 62.23%, 64.81%, and 65.54%, respectively. It is seen from the results that although higher associativity can get higher hit rates, it adversely impacts the reliability.

5.2. Cache line size

When the processor accesses a part of memory that is not already in the cache, it loads a chunk of the memory around the accessed address into the cache, hoping that it will soon be used again. When data are transferred between cache and main memory, this chunk of data is handled in a fixed size, called cache lines. A cache can only hold a limited number of lines, determined by the cache size. For example, a 64 KB cache with 64-byte lines has 1024 cache lines. In LEON3, cache line size can be configured as 16-byte or 32-byte. Other configurations are kept the same: two-way set associative, 32 KB cache size, and LRU replacement algorithm.

Figure 12 shows the failure probabilities for 16-byte and 32-byte cache line size for the six tested benchmarks. It is obviously seen that, for all the tested benchmarks, 32-byte cache line has lower failure probability than 16-byte, meaning 32-byte configuration is more reliable than 16-byte. Besides, 32-byte also achieves better hit rates than 16-byte for four of the six

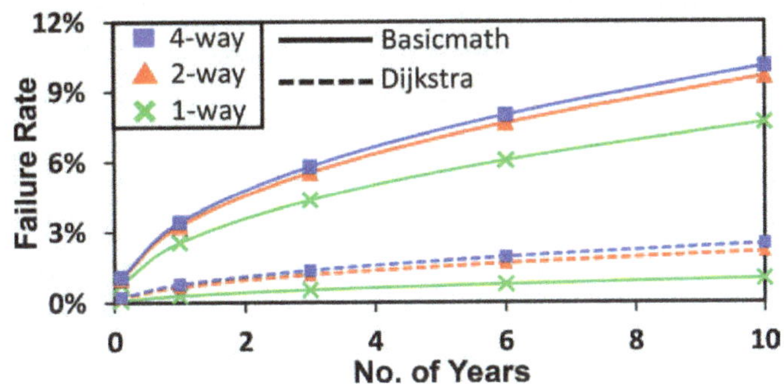

Figure 11. The failure probability as a function of time for three different associativities and two benchmarks.

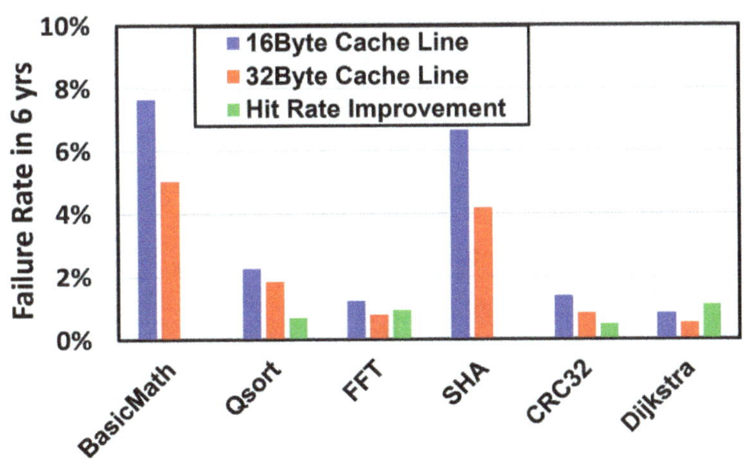

Figure 12. The failure probabilities in 6 years for 16-byte cache line and 32-byte cache line for six applications. The hit-rate improvement is also shown, defined as the improvement of using 32-byte cache line compared to 16-byte line.

benchmarks except for SHA and Basicmath, and hit rates for 32-byte and 16-byte are almost the same. Overall, from our observation, larger cache line size can improve both hit rate and reliability.

The reason for that is, a cache miss in a 32 Byte cache line can produce more recovery cycles up to 256 (32 × 8) SRAM cells, which is twice as with a 16-byte cache line (16 × 8 SRAM cells). The more BTI recovery cycles, the better reliability the cache would have.

5.3. Cache size

In our experiments, we have set five different cache sizes for the data cache of LEON3: 4, 16, 32, 64, and 128 KB. Other configurations are kept the same: two-way set associative, 16-byte cache linesize, and LRU replacement algorithm. In **Figure 13**, the hit rate and probability that the data cache fails in 6 years are presented for different cache sizes. As expected, the larger the cache size, the cache is more vulnerable and less reliable. For hit rate, although larger cache size always results in better hit rates, the improvement is little when cache size is larger than 32 KB. It is also worth noting that larger cache size causes more area and more power.

5.4. Replacement algorithm

If all the cache lines in the cache are in use, when the microprocessor accesses a new line, one of the lines currently in the cache must be evicted to make room for the new line. The policy that the microprocessor uses to choose the entry to evict is called the replacement policy.

The heuristic of any replacement policy is that it tries to predict which existing entry is the least likely to be used in the future. The most common replacement policy in modern processors is least recently used (LRU) policy. The Least-Recent-Replaced (LRR) algorithm evicts the cache entry, which is least recently replaced. Another replacement policy is random replacement, meaning that a random cache line is selected for eviction. Among them, random

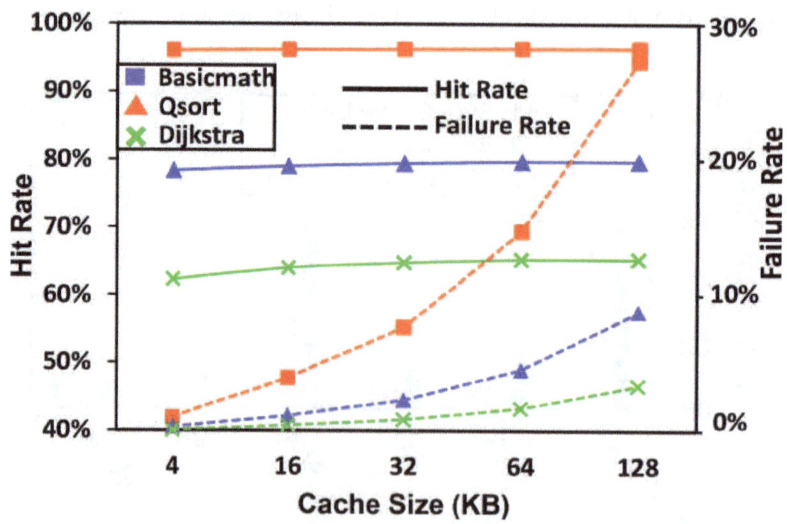

Figure 13. The hit rate and the failure probability in 6 years are shown for five different cache sizes and for three applications.

replacement policy is the simplest. It has low area overhead but suffers from poor cache efficiency. LRR algorithm uses one extra bit in the tag part, and it also has low area overhead. LRU algorithm typically has the best performance but with the cost of the highest area overhead among the three.

In this chapter, we have configured LEON3 to three different replacement algorithms, LRR, LRU, and Random. Other configurations are kept the same: two-way set associative, 16-byte cache line size, and 32 KB cache size.

Figure 14 shows the failure probabilities for the three replacement algorithms as well as the hit-rate improvement of LRU and LRR compared to Random. As expected, LRU has the best hit rate for all the tested benchmarks. However, seen from the results, it has lower reliability compared to LRR and Random. The reason for the abovementioned results is LRU has better hit rate and fewer misses, which result in fewer recovery cycles.

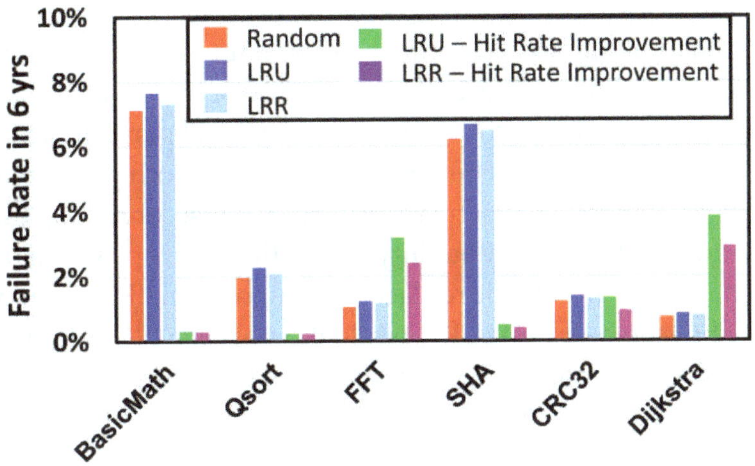

Figure 14. The failure probabilities in 6 years for 16-byte cache line and 32-byte cache line for six applications. The hit-rate improvement is also shown, defined as the improvement of using 32-byte cache line compared to 16-byte line.

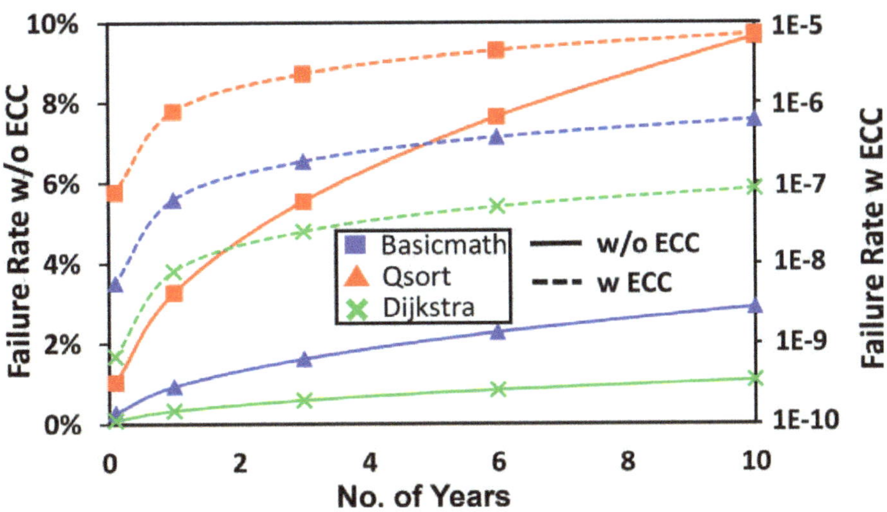

Figure 15. The failure probabilities of the two-way 32 KB data cache with and without ECC codes are shown as a function of time for three applications.

5.5. Error correcting codes

Error correcting codes (ECC) is used to detect and correct internal data corruptions in SRAMs. It uses some extra bits to check the data consistency and to correct the corrupted data. As mentioned, BCH codes [37] was used which consumes seven additional bits per word and can correct one bit per word, meaning the number of bits per word is 39 with the inclusion of ECC for LEON3.

Figure 15 shows the failure probabilities of the data cache for with and without ECC. Again, the failure probabilities are a function of time. Three illustrative benchmarks are present (other benchmarks have similar results). As shown in the results, ECC can significantly improve cache reliability.

6. Insights and conclusions

We have shown the reliability and performance of the data cache for different configurations. For associativity, larger associativity has better performance but worse reliability. According to the results, two-way set associative cache achieves the optimal performance-reliability balance. For cache line size, 32-byte cache line is better than 16-byte in both performance and reliability. Cache size is of great significance to cache reliability. We also observed that when cache size increases larger than 16 KB, the cache reliability dramatically drops while the performance (hit rate) has very limited improvement. For replacement algorithm, 'Random' replacement policy has the worst hit rate but the best reliability, while the popular LRU algorithm has the best hit rate but the worst reliability among the three. Therefore, tradeoffs can be made between the three replacement algorithms. ECC always improves reliability with area and power overhead.

Overall, experimental results show that the cache size and ECC codes are of great significance to cache reliability, while other metrics have smaller impact. According to the performance-reliability evaluation, an optimal tradeoff could be achieved for the cache design in a microprocessor system.

Author details

Taizhi Liu*, Chang-Chih Chen and Linda Milor

*Address all correspondence to: taizhiliu88@gatech.edu

Georgia Institute of Technology, Georgia

References

[1] Jouppi NP et al. Improving direct-mapped cache performance by the addition of a small fully-associative cache and prefetch buffers. In: Proceedings of 17th International Symposium on Computer Architecture (ISCA-17). 1990. pp. 364-373

[2] Albonesi DH. Selective cache ways: On-demand cache resource allocation. In: Proceedings of 32nd International Symposium on Microarchiteure (MICRO-32). 1999. pp. 248-259

[3] Jaleel A et al. High performance cache replacement using re-reference interval prediction (RRIP). In: Proceedings of 37th International Symposium on Computer Architecture (ISCA-37). 2010. pp. 60-71

[4] Kaczer B et al. Atomistic approach to variability of bias-temperature instability in circuit simulations. In: Proceedings of IEEE International Reliability Physics Symposium (IRPS). 2011. pp. XT.3.1-XT.3.5

[5] Huard V et al. NBTI degradation: From transistor to SRAM arrays. In: Proceedings of IEEE International Reliability Physics Symposium (IRPS). 2008. pp. 289-300

[6] Bansal A et al. Impact of NBTI and PBTI in SRAM bit-cells: Relative sensitivities and guidelines for application-specific target stability/performance. In: Proceedings of IEEE International Reliability Physics Symposium (IRPS). 2009. pp. 745-749

[7] Lin JC et al. Time dependent Vccmin degradation of SRAM fabricated with high-k gate Dielectris. In: Proceedings of IEEE International Reliability Physics Symposium (IRPS). 2007. pp. 439-444

[8] Kang K et al. Impact of negative-bias temperature instability in nanoscale SRAM array: Modeling and analysis. In: TCAD. 2007. pp. 1770-1781

[9] Bansal A et al. Impacts of NBTI and PBTI on SRAM static/dynamic noise margins and cell failure probability. Journal Microelectronics and reliability. 2009;**49**:642-649

[10] Bansal A, Kim J-J, Rao R. Usage-based degradation of SRAM arrays due to bias temperature instability. In: Proceedings of IEEE International Reliability Physics Symposium (IRPS). 2012. pp. 2F.6.1-2F.6.4

[11] Weckx P et al. Defect-based methodology for workload-dependent circuit lifetime projections-Application to SRAM. In: Proceedings of IEEE International Reliability Physics Symposium (IRPS). 2013. pp. 3A.4.1-3A.4.7

[12] Angot D et al. The impact of high Vth drifts tail and real workloads on SRAM reliability. In: Proceedings of IEEE International Reliability Physics Symposium (IRPS). 2014. pp. CA.10.1-CA.10.6

[13] Mintarno E et al. Workload dependent NBTI and PBTI analysis for a sub-45 nm commercial microprocessor. In: Proceedings of IEEE International Reliability Physics Symposium (IRPS). 2013. pp. 3A.1.1-3A.1.6

[14] Khan S et al. Trends and challenges of SRAM reliability in the nano-scale era. In: Proceedings of Design and Technology of Integrated Systems in Nanoscale Era (DTIS). 2010. pp. 1-6

[15] Indaco M et al. On the impact of process variability and aging on the reliability of emerging memories (embedded tutorial). In: Proceedings of European Test Symposium (ETS). 2014. pp. 1-10

[16] Huard V et al. Managing SRAM reliability from bitcell to library level. In: Proceedings of IEEE International Reliability Physics Symposium (IRPS). 2010. pp. 655-664

[17] Qin J et al. SRAM stability analysis considering gate oxide SBD, NBTI and HCI. In: Proceedings of International Integrated Reliability Workshop (IIRW). 2007. pp. 33-37

[18] Siddiqua T et al. Recovery boosting: A technique to enhance NBTI recovery in SRAM arrays. In: Proceedings of IEEE Computer Society Annual Symposium VLSI. 2010. pp. 393-398

[19] Gunadi E et al. Combating aging with the colt duty cycle equalizer. In: Proceedings of 43rd International Symposium on Microarchiteure (MICRO-43). 2010. pp. 103-114

[20] Shin J et al. A proactive Wearout recovery approach for exploiting microarchitectural redundancy to extend cache SRAM lifetime. In: Proceedings of 35th International Symposium on Computer Architecture (ISCA-35). 2008. pp. 353-362

[21] Wirth GI, da Silva R, Kaczer B. Statistical model for MOSFET bias temperature instability component due to charge trapping. IEEE Transactions on Electron Devices. 2011;**58**(8):2743-2751

[22] Fernandez R, Kaczer B, Nackaerts A, Demuynck S, Rodriguez R, Nafria M, Groeseneken G. AC NBTI studies in the 1 Hz–2 GHz range on dedicated on-chip CMOS circuit. In: Proceedings of International Electron Devices Meeting. 2006

[23] Zafar S, Kim YH, Narayanan V, Cabral C, Paruchuri V, Doris B, Stathis J, Callegari A, Chudzik M. A comparative study of NBTI and PBTI (charge trapping) in S_iO_2/HFO_2 stacks with FUSI, TiN, Re Gates. In: Proceedings of Symposium VLSI Technology. 2006. pp. 23-25

[24] Chen S-Y, Tu C-H, Kao P-W, Lin M-H, Haung H-S, Lin J-C, Wang M-C, Wu S-H, Jhou Z-W, Chou S, Ko J. Investigation of DC hot-carrier degradation at elevated temperatures for p-channel metal-oxide-semiconductor field-effect transistors. Japanese Journal of Applied Physics. 2008;**47**(3):1527-1531

[25] Wang W, Reddy V, Krishnan AT, Vattikonda R, Krishnan S, Cao Y. Compact modeling and simulation of circuit reliability for 65-nm CMOS technology. IEEE Transaction on Device and Materials Reliability. 2007;**7**(4):509-517

[26] Liu T, Chen C-C, Cha S, Milor L. System-level variation-aware aging simulator using a unified novel gate-delay model for bias temperature instability, hot carrier injection, and gate oxide breakdown. Microelectronics Reliability. 2015. DOI: 10.1016/j.microrel.2015.06.008.

[27] Liu T, Chen C-C, Kim W, Milor L. Comprehensive reliability and aging analysis on SRAMs within microprocessor systems. Microelectronics Reliability. 2015. DOI: 10.1016/j.microrel.2015.06.078

[28] Ma C, Li B, Zhang L, He J, Zhang X, Lin X, Chan M. A unified FinFET reliability model including high K gate stack dynamic threshold voltage, hot carrier injection, and negative bias temperature instability. In: Proceedings of International Symposium Quality Electronic Design (ISQED). 2009. pp. 7-12

[29] Tu CH, Chen SY, Chuang AE, Huang HS, Jhou ZW, Chang CJ, Chou S, Ko J. Transistor variability after CHC and NBTI stress in 90 nm pMOSFET technology. Electronics Letters. 2009;**45**(15):854-856

[30] Calimera A et al. Partitioned cache architectures for reduced NBTI-induced aging. In: Proceedings of DATE. 2011. pp. 1-6

[31] PrimeTime Power Modeling Tool. [Online]. Available: http://www.synopsys.com/Tools/Implementation/SignOff/PrimeTime/Pages/default.aspx [Accessed June, 2014]

[32] HotSpot Temperature Modeling Tool. [Online]. Available: http://lava.cs.virginia.edu/HotSpot [Accessed March, 2014]

[33] LEON3 Processor. Available: http://gaisler.com/index.php/downloads/leongrlib [Accessed December, 2015]

[34] Mibench benchmark: http://www.eecs.umich.edu/mibench

[35] Seevinck E, List FJ, Lohstroh J. Static-noise margin analysis of MOS SRAM cells. IEEE Journal of Solid-State Circuits. 1987;**22**(5):748-754

[36] Kang R, Joshi R, Nassif S. Mixture importance sampling and its application to the analysis of SRAM designs in the presence of rare failure events. In: Proceedings of Design Automation Conference. 2006. pp. 69-72

[37] Sklar B, Harris FJ. The ABCs of linear block codes. IEEE Signal Processing Magazine. 2004;**21**:14-35

[38] Ricketts A et al. Investigating the impact of NBTI on different power saving cache strategies. In: Proceedings of DATE. 2010. pp. 592-597

[39] Pekhimenko G et al. Base-delta-immediate compression: practical data compression for on-chip caches. In: Proceedings of of 21st International Conference on Parallel Architecture and Compilation Techniques (PACT-12). 2012. pp. 377-388

11

Stochastic Reward Net-based Modeling Approach for Availability Quantification of Data Center Systems

Tuan Anh Nguyen, Dugki Min and Eunmi Choi

Abstract

Availability quantification and prediction of IT infrastructure in data centers are of paramount importance for online business enterprises. In this chapter, we present comprehensive availability models for practical case studies in order to demonstrate a state-space stochastic reward net model for typical data center systems for quantitative assessment of system availability. We present stochastic reward net models of a virtualized server system, a data center network based on DCell topology, and a conceptual data center for disaster tolerance. The systems are then evaluated against various metrics of interest, including steady state availability, downtime and downtime cost, and sensitivity analysis.

Keywords: virtualized servers system, data center system, disaster tolerant data center

1. Introduction

Data centers (DCs) have been the core-centric of modern ICT ecosystems in recent decades. Computing resources and crucial telecommunications are centralized in a data center to constantly facilitate online business and to connect people from distant parts of the world through the internet. Giant internet companies such as Facebook, Amazon, and Google have built huge state-of-the-art centers to house their own IT infrastructure. According to a study by the Ponemon Institute [1] regarding the cost of data center outages from 63 DCs located in the United States over a 12-month period, the average cost due to unplanned outages in 2016 was US$ 740,357, which steadily increased by 46% from US$ 505,502 since it was first studied in 2010. Specifically, a minute of downtime costs around US$ 7900 on average. However, online businesses actually face more severe revenue losses due to IT service downtime. In early 2016, Amazon suffered an incredible business loss of US$ 66,240/minute due to server downtime over a period of approximately 15 minutes. The causes of system outages in DCs span from uncertain failures of IT parts/blocks to natural disasters. Therefore, a quantification of IT infrastructure availability in DCs under various scenarios in advance of system development is of paramount importance for big tech companies.

Availability assessment approaches are primarily based on measurement and modeling methods. Model-based approaches are fast and relatively inexpensive methods for system availability analysis in comparison with measurement-based methods. System modeling can be accomplished using discrete-event simulation [2, 3], analytical models, or a hybrid of both approaches. Analytical models fall into four main categories [4–7]: (i) non-state-space models (reliability graph (RelGraph), reliability block diagram (RBG), or fault tree (FT)), state-space models (Markov chains, Stochastic Petri net (SPN), stochastic reward net (SRN), etc.), hierarchical models, and fixed-point iterative models. Non-state-space modeling paradigms provide a relatively quick evaluation of basic metrics for a system (reliability, availability, MTTF) with a proper capture of overall system architecture. State-space models, on the other hand, can capture sophisticated behaviors and operations of a system. This approach can handle failure/repair dependencies and complex interactions between system components. To avoid the largeness problem (or state-space explosion problem) in state-space models, we use hierarchical modeling techniques of non-state-space and state-space models at upper and lower levels, as well as fix-point iterative models. In this chapter, we focus on studying complex system operations in DCs captured by using an SRN.

The structure of this chapter is organized into six sections. Section 2 provides preliminary concepts of availability modeling and analysis of data center systems (DCS). Subsequently, several case studies are presented. Section 3 offers an availability model of a unit system of the virtualized server (VSS) in DCs. In Section 4, we present availability modeling of a data center network (DCN) based on DCell topology. We present an SRN model for a DC in order to study disaster tolerance in Section 5. Finally, we present conclusions in Section 6.

2. Availability quantification of data center systems: basic concepts

Availability A(t) of a DCS represents the probability of its operating system taking the correct state at an instant t, regardless of the number of failures and repairs during the interval $(0,t)$. *Instantaneous/point* availability $A(t)$ is related to the system reliability, as defined in Eq. (1).

$$A(t) = R(t) + \int_0^t R(t-x)g(x)dx \tag{1}$$

$R(t)$ is the instantaneous reliability at t of the system, which is defined in Eq. (2):

$$R(t) = \int_t^\infty f(x)dx \tag{2}$$

$f(x)$ is the probability density function of a random variable X, which represents the system's lifetime or time to failure.

$g(x)$ is a renewal process rate in the interval $(0,t)$, as defined in Eq. (3)

$$g(x) = f(x) + \int_0^x g(x-u)f(u)du \tag{3}$$

$m(x)dx$ is the probability that a renewal process cycle will be completed in the time interval $[x, x+dx]$. $R(t-x)$ is the probability that the system works properly for the remaining time interval $t-x$. $R(t-x)m(x)dx$ is the probability of the case that a fault has occurred and that after the repair/renewal (which occurred at the instance x, $0<x<t$), the system resumed functioning with no further faults. If a system is not repairable, the concept of $A(t)$ is identical with that of reliability $R(t)$.

Steady-state availability (SSA) is the system availability after a long running time, where the limiting value $A(t)$ tends to decrease from 1 at the initial instant, as defined in Eq. (4) and Eq. (5).

$$A = \lim_{t\to\infty} A(t) = \frac{MTTF}{MTTF + MTTR} \tag{4}$$

$$A = \lim_{t\to\infty} A(t) = \frac{\mu}{\lambda + \mu} \tag{5}$$

The *failure rate* (λ) implies the frequency of system failure is determined by the total number of failures within an item population, divided by the total time expended by that population, during a particular measurement interval under the stated conditions. *Repair rate* (μ) implies the frequency of system repair determined as the average number of repairs over a period of maintenance time. *Mean time to failure (MTTF)* represents the expected time in which a system functions correctly before its first failure. *Mean time to repair (MTTR)* represents the expected time required for system repair. In the case where failure/repair events comply with exponential distributions, *MTTF* and *MTTR* represent an arithmetic inversion of failure and repair rates, as shown in Eq. (6). SSA can be computed from Eq. (5).

$$MTTF = \frac{1}{\lambda} \quad MTTR = \frac{1}{\mu} \tag{6}$$

In industry, system administrators are usually concerned with *system downtime* (measured in minutes per year) and *downtime cost* (with a cost unit C per minute of system downtime). These values can be computed with Eq. (7) and (8).

$$Downtime = (1 - A)*8760^{*}60 \tag{7}$$

$$Downtime\ \mathrm{Cos}\,t = C^{*}(1 - A)*8760^{*}60 \tag{8}$$

Sensitivity analysis is performed to assess the importance of system parameters by two techniques. (*i*) Repeatedly substitute specific parameter values in one range at a time while the others remain constant, and observe system behaviors in accordance with the variation of the selected parameter. This approach studies the system responses upon a broad range of the parameters under consideration. (*ii*) *Differential sensitivity analysis*: compute partial derivatives of the measure of interest with respect to each system parameter as determined in Eq. (9) or (10) to yield a scaled sensitivity.

$$S_\tau(A) = \frac{\partial A}{\partial \tau} \tag{9}$$

$$SS_\tau(A) = \frac{\partial A}{\partial \tau}\left(\frac{\tau}{A}\right) \tag{10}$$

Stochastic reward net (SRN) [8] has been an appropriate modeling paradigm to capture operational complexities in industrial hardware and software systems [9–14]. According to a specific description of system operations, ones can model system behaviors using place(s), transition(s) and arc(s) as three main components in an SRN model. To represent a certain entity of the system to be considered, we use token(s) (normally denoted by a dot or an integer number to represent a number of corresponding entities) which reside in each place of the SRN model. And to capture its operational state variations, we use (input/output) arcs to connect transition(s) to place(s) or place(s) to transition(s), respectively. A firing of a transition is triggered when a certain condition of system state is matched in order to allow the token(s) in a place are removed, and then deposited in another place. The transitions of tokens in an SRN model captures the system's operations while the residence of tokens in places represent the system's operational state at a time, which is call marking. The Boolean condition attached to each transition which is to enable/disable the transition is called the guard. A set of guard functions can be defined to articulate the behaviors of system state dependence and transition. A marking-dependence (denoted by a # sign attached to a transition) is incorporated when the transition's rate is dependent on the marking of the SRN model at a time. Other features of SRN including inhibitor arcs, multiplicities, and input arcs can simplify the construction of SRN models.

SRN-based availability quantification framework is presented in **Figure 1**. The availability quantification framework consists of three stages: (i) requirement specification, (ii) SRN-based system modeling and (iii) system analysis. Service level agreement (SLA) [15, 16] between system owner and customer details system specification and requirements. In the stage (i), taking into

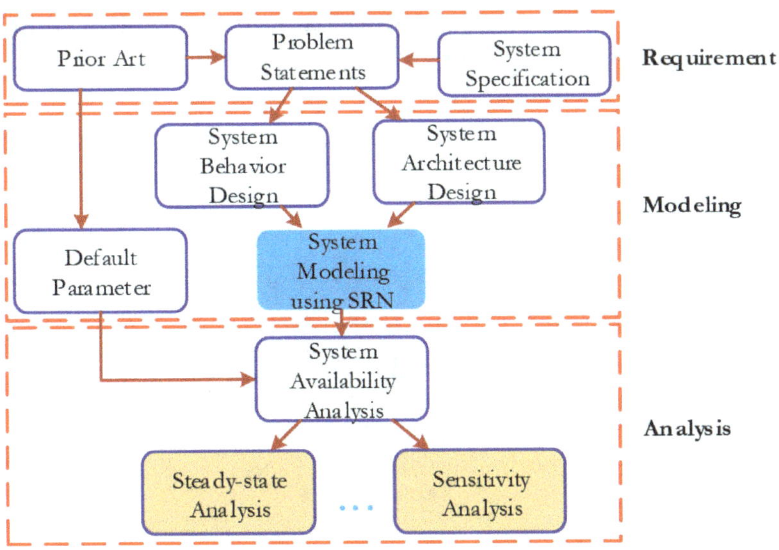

Figure 1. SRN-based availability quantification framework.

account the literature review based on prior art and contemporary development of the system, ones can define problem statements to be modeled and observed. In the stage (ii), the person in charge of modeling and evaluating the system can refer various default values of system parameters from previous work. He/she can propose the architecture design and detailed behaviors taken into consideration of the system. The SRN is used to capture the pre-defined system operations. The SRN system model is then analyzed and the system availability evaluation is performed with regard to various output measures of interest via different analysis approaches such as steady-state availability and/or sensitivity analysis.

3. Case study I: a virtualized server system

3.1. System architecture

Figure 2 shows a general VSS architecture. A VSS is a computing unit in a DC which consists of a number of physical servers (also called hosts *H1, H2, …,* H_n). Each server is in turn virtualized using bare-metal virtualization technology [17–19]. Thus, each server hosts its own hypervisor (hereinafter, called the virtual machine monitor (VMM)). The physical server is capable of running a number of virtual machines (VM) on top of its VMM. For the sake of fault tolerance and data storage of VMMs and VMs, the physical servers are interconnected via a network pipeline to each other, and to a shared storage area network (SAN).

To focus on modeling complex behaviors of a virtualized system in a detailed manner, we consider a small-size VSS consisting of two hosts (*H1* and *H2*) connected to a shared SAN. Each host runs its own virtual machine monitors *VMM1* and *VMM2*, respectively. Two VMs are also created on each host, *VM1* for host *H1* and *VM2* for host *H2*. In the next section, we will present SRN models of the above-mentioned subsystems. The models capture in detail various failure modes and recovery methods, including hardware failures in physical hosts and SAN [20, 21], failures due to non-aging related Mandelbugs on both VMM and VM subsystems [22], and software aging-related failures and corresponding time-interval software rejuvenation techniques for VMM and VM subsystems [23, 24]. Furthermore, we incorporate

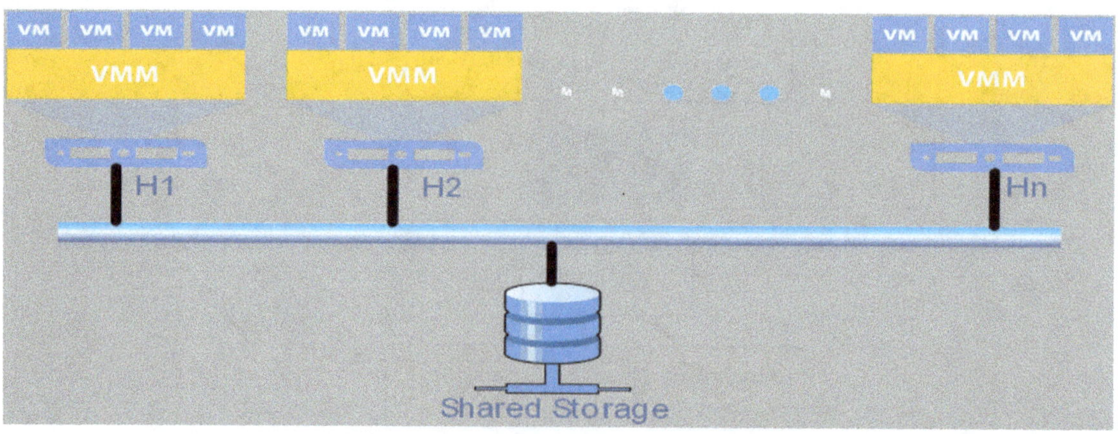

Figure 2. A virtualized server system with two physical servers.

hierarchically complex dependencies between subsystems, including the dependences of a VM on its VMM, a VM on the shared SAN, and a VMM on its host. Without loss of generality, the proposed SRN model represents the sophisticated operations of, and interactions between subsystems, in a typical virtualized system as a computing unit brick in a practical DC. The model can be further extended in the future by incorporating a large scale cloud system as in [25].

3.2. SRN models of VSS

The SRN system model is presented in **Figure 3**. We use a two-state SRN model to capture the operational state (UP) and failed state (DOWN) of the physical parts, including host 1 (*H1*), host 2 (*H2*), and SAN, as shown in **Figure 3(a)–(c)**, respectively.

The VMM subsystem models are shown in **Figure 3(d)** and **(f)** for *VMM1* and its clock, respectively, and in **Figure 3(e)** and **(g)** for *VMM2* and its clock, respectively. Without loss of generality, a model of a VMM (either *VMM1* or *VMM2*) subsystem consists of six states (represented by shaded places): (i) normally running state (P_{VMMup}), (ii) failure state due to non-Mandelbugs (P_{VMMf}), (iii) down-state due to a failure of its underlying host (P_{VMMdn}), (iv) failure-probable state due to aging problems (P_{VMMfp}), (v) aging-failure state due to aging of equipment (P_{VMMaf}), and (vi) rejuvenation-process state (P_{VMMrej}). Initially, there is a token in P_{VMMup} to represent a running VMM. If it fails due to a non-aging Mandelbug, the transition time T_{VMMf} is fired to transit the token into P_{VMMf}. Recovery is captured by $T_{VMMrepair}$. After running for a long time, the VMM suffers a high failure probability while remaining operational. Therefore, it goes to the failure-probable state P_{VMMfp} as T_{VMMfp} is fired. Failure due to aging occurs soon after T_{VMMaf} is fired and the VMM goes to the aging-failure state P_{VMMaf}. Its recovery is represented by the firing of T_{VMMar}. If the VMM's underlying host goes down (i.e., a token is deposited in P_{Hf} in respective **Figure 3(a)** or **(b)**) while the VMM is in the UP states (normal P_{VMMup} or failure-probable P_{VMMfp}), the VMM immediately enters the down-state P_{VMMdn} through the immediate fired transitions $t_{VMMupdn}$ or $t_{VMMfpdn}$. A reset is necessary for the VMM to go up (captured by $T_{VMMreset}$) after its host is recovered. In the meantime, the VMM clock is initiated by a token in $P_{VMMclock}$, which counts time by firing a timed transition $T_{VMMclockinterval}$ that complies with the c_{VMM}-stage Erlang distribution. Every software rejuvenation process interval on a VMM is represented by a firing of $T_{VMMclockinterval}$, and the token in $P_{VMMclock}$ is removed and deposited in $P_{VMMpolicy}$. Thus, rejuvenation is triggered if there is a VMM in P_{VMMup} or P_{VMMfp} by firing the immediate transitions $t_{VMMuprej}$ or t_{VMMrej}. Also, the token in $P_{VMMpolicy}$ of the VMM clock model is moved to $P_{VMMtrigger}$. The VMM represented by a token in P_{VMMrej} is then rejuvenated and returned to the normal state P_{VMMup} as T_{VMMrej} is fired. The VMM clock is reset as $t_{VMMclockreset}$ is fired to start a new interval of time-based software rejuvenation on a VMM. The modeling of VMM1 on host H1 and VMM2 on host H2 are identical based on the general model description as above.

Modeling of VM subsystems is shown in **Figure 3(h)** and **(j)** for VM1 subsystem and its clock, respectively, and **Figure 3(i)** and **(k)** for VM2 subsystem and its clock, respectively. The models initiate with two tokens in P_{VMup} representing two VMs on each host. In general, the SRN model of a VM subsystem also consists of six states as in the VMM subsystem does including: (i) normal state (P_{VMup}), (ii) failure state due to non-aging Mandelbugs (P_{VMf}), (iii) down-state

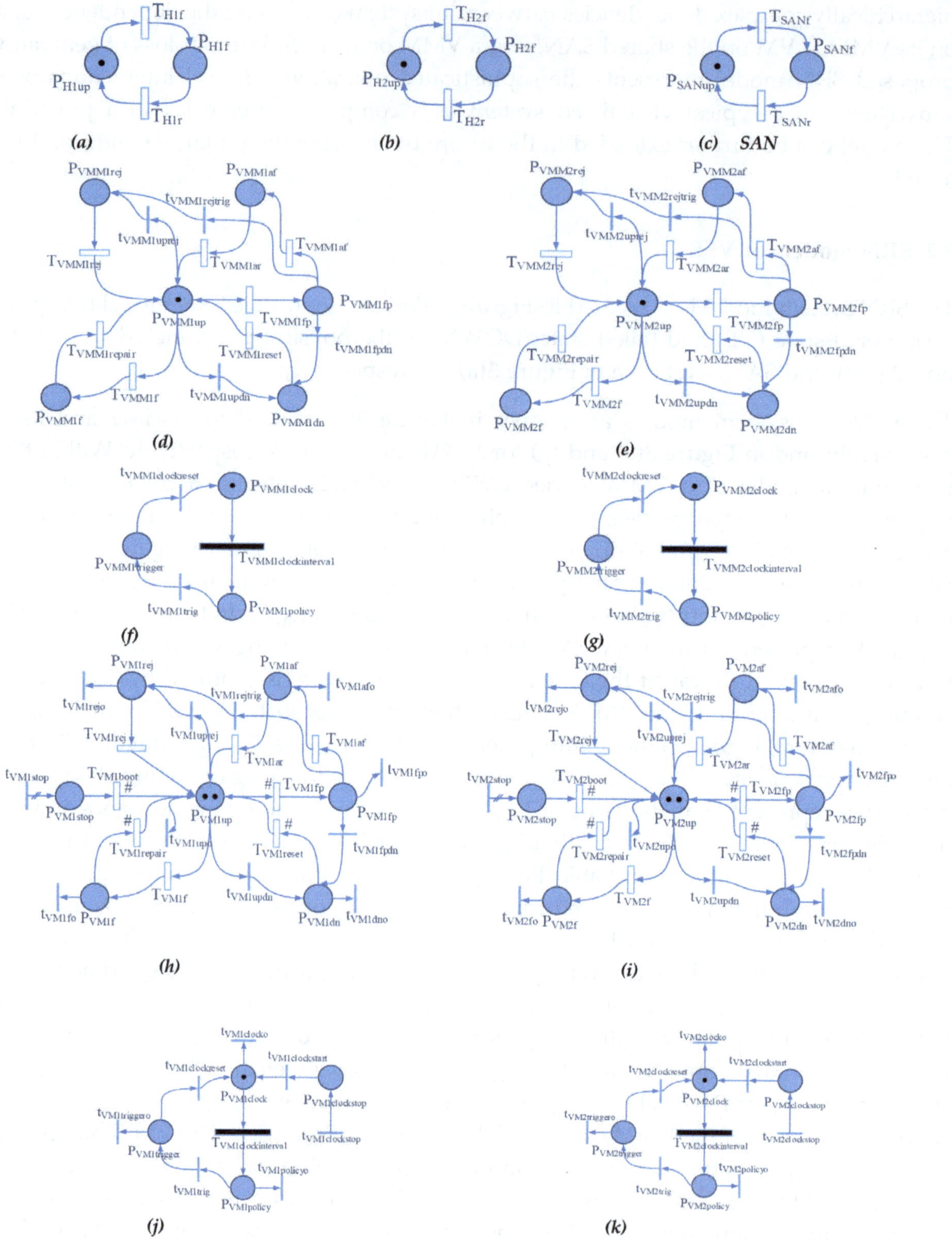

Figure 3. SRN system model of a VSS: (a) Host 1, (b) Host 2, (c) SAN, (d) VMM1, (e) VMM2, (f) VMM1's clock, (g) VMM2's clock, (h) VM1, (i) VM2, (j) VM1's clock, and (k) VM2's clock.

due to a failure of underlying VMM (P_{VMdn}), (iv) failure-probable state due to aging problems (P_{VMfp}), (v) aging-failure state due to a failure of aging (P_{VMaf}) and (vi) rejuvenation-process state (P_{VMrej}). The operations of the VM subsystem in correspondence with the transitions of tokens in the SRN model are similarly described as those of the VMM subsystem. However, the SRN model of the VM subsystem is further extended by incorporating (i) marking-dependence represented by a "#" mark nearby selected timed transitions (T_{VMfp}, T_{VMf}, $T_{VMreset}$) to capture the cases in which two VMs in the same state compete with each other in order to transit to a new state and (ii) dependence between the VM subsystem and SAN. The second dependence is captured by the immediate transitions t_{VMupo}, t_{VMfo}, t_{VMdno}, t_{VMfpo}, t_{VMafo}, and t_{VMrejo} in the VM model, and $t_{VMclocko}$, $t_{VMpolicyo}$, and $t_{VMtriggero}$ in the VM clock model. As the SAN fails (depicted by a token in P_{SANf}), these transitions are fired to remove tokens in the VM model and VM clock model, regardless of their locations representing the loss of VM images on SAN and VM clock functionalities. Nevertheless, as soon as the SAN is recovered, two VMs are immediately created on the SAN, and they are booted onto a VMM of a corresponding host. The creation of multiple VMs is captured by t_{VMstop}, whereas the booting of a VM in the sequence is captured by T_{VMboot} with marking-dependence. The VM clock is also started after the recovery of a SAN, as captured by $P_{VMclockstop}$ and two immediate transitions $t_{VMclockstop}$ and $t_{VMclockstart}$.

3.3. Availability analysis scenarios and results

We implemented the SRN models in the Stochastic Petri Net Package (SPNP) [26]. Input parameters are selected based on previous work [20, 27], as shown in **Table 1**.

Input	Description	Transitions	Value	Input	Description	Transitions	Value
μ_{hr}	Host repair	T_{H1r}, T_{H2r}	3 days	λ_{hf}	Host fail	T_{H1f}, T_{H2f}	1 years
λ_{vmmf}	VMM non-aging failure	T_{VMM1f}, T_{VMM2f}	2654 hours	λ_{vmf}	VM non-aging failure	T_{VM1f}, T_{VM2f}	2893 hours
μ_{vmmr}	VMM reset	$T_{VMM1reset}$, $T_{VMM2reset}$	1 min	δ_{vmr}	VM repair	$T_{VM1repair}$, $T_{VM2repair}$	30 min
δ_{vmmr}	VMM repair	$T_{VMM1repair}$, $T_{VMM2repair}$	100 min	μ_{vmr}	VM restart	$T_{VM1reset}$, $T_{VM2reset}$	50s
β_{vmmfp}	VMM failure-probable	T_{VMM1fp}, T_{VMM2fp}	2 months	β_{vmfp}	VM failure-probable	T_{VM1fp}, T_{VM2fp}	1 month
λ_{vmmaf}	VMM aging-failure	T_{VMM1af}, T_{VMM2af}	2 weeks	λ_{vmaf}	VM aging failure	T_{VM1af}, T_{VM2af}	1 week
μ_{vmmar}	VMM aging recovery	T_{VMM1ar}, T_{VMM2ar}	120 min	μ_{vmar}	VM aging recovery	T_{VM1ar}, T_{VM2ar}	120 min
τ_{vmm}	VMM clock interval	$T_{VMM1clockinterval}$, $T_{VMM1clockinterval}$	1 week	τ_{vm}	VM clock interval	$T_{VM1clockinterval}$, $T_{VM2clockinterval}$	3.5 days
β_{vmmrej}	VMM rejuvenation	$T_{VMM1rej}$, $T_{VMM1rej}$	2 min	β_{vmrej}	VM rejuvenation	T_{VM1rej}, T_{VM2rej}	1 min
λ_{sf} μ_{sr}	SAN fail SAN repair	T_{SANf} $T_{SANrepair}$	1 year 3 days	η_{vmb}	VM booting after VMM rejuvenation	$T_{VM1boot}$, $T_{VM2boot}$	50s
c_{VMM}	c_{VMM}-stage Erlang distribution	x	10	c_{VM}	c_{VM}-stage Erlang distribution	X	10

Table 1. Input parameters of SRN models.

Cases	Description	SSA of VMM	SSA of VM
I	Rejuvenation is applied on all VMM and VM subsystems in both hosts.	0.999912470996	0.991769547666
II	Rejuvenation is not applied only on one of VMM subsystems in two hosts but applied on both VM subsystems in two hosts.	0.999908948744	0.991766082049
III	Rejuvenation is applied on both VMM subsystems in two hosts but not applied to only one of two VM subsystems.	0.999912470996	0.991770317258
IV	Rejuvenation is not applied on haft side of the system including VMM1 and VM1 subsystems but applied on VMM2 and VM2 subsystems.	0.999908948744	0.991766912872
V	Rejuvenation is not applied on both VMM subsystems in two hosts but applied on both VM subsystems.	0.999905284754	0.991763344539
VI	Rejuvenation is applied on both VMM subsystems in two hosts, but not applied on both VM subsystems.	0.999912470996	0.991771080172
VII	Rejuvenation is not applied on VMM and VM subsystems in both hosts.	0.999905284754	0.99176419998

Table 2. Analysis scenarios of VSS and SSAs of VMM and VM subsystems.

- *Steady-state availability:* We conducted numerical experiments in seven case studies with regard to different rejuvenation combinations. The case studies are described along with analysis results of SSA of VMM and SSA of VM in **Table 2**. The reward functions used to compute SSAs are defined as

$$SSA_{VMM} = \begin{cases} 1 : if\ \left(\#P_{VMM1up} + \#P_{VMM1fp} + \#P_{VMM2up} + \#P_{VMM2fp}\right) > 0 \\ 0 : otherwise \end{cases}$$
$$SSA_{VM} = \begin{cases} 1 : if\ \left(\#P_{VM1up} + \#P_{VM1fp} + \#P_{VM2up} + \#P_{VM2fp}\right) > 0 \\ 0 : otherwise \end{cases} \quad (11)$$

where $\#P_X$ is the number of token in place P_X. The results show that the following:

i. Time-based rejuvenation techniques with default parameters, when implemented on both VMM and VM subsystems in combination does not gain the highest SSA for the virtualized system. When a VMM undergoes a rejuvenation process, it pulls down all VMs running on top of the VMM;

ii. Rejuvenation on VMM exposes more effectiveness in gaining higher SSA in comparison to the VM.

iii. An appropriate rejuvenation combination implemented on either a VMM or VM with proper clock intervals can actually enhance system availability.

- *Sensitivity analysis of SSA*: The sensitivity analysis is observed in five case studies w.r.t the variation of: (i) only VMM1 clock's interval; (ii) only VM1 clock's interval; (iii) both VMM1 and VMM2 clocks' interval; (iv) both VM1 and VM2 clocks' interval; and (v) all clock intervals with the same duration, as shown in **Figure 4**. The findings are as follows:

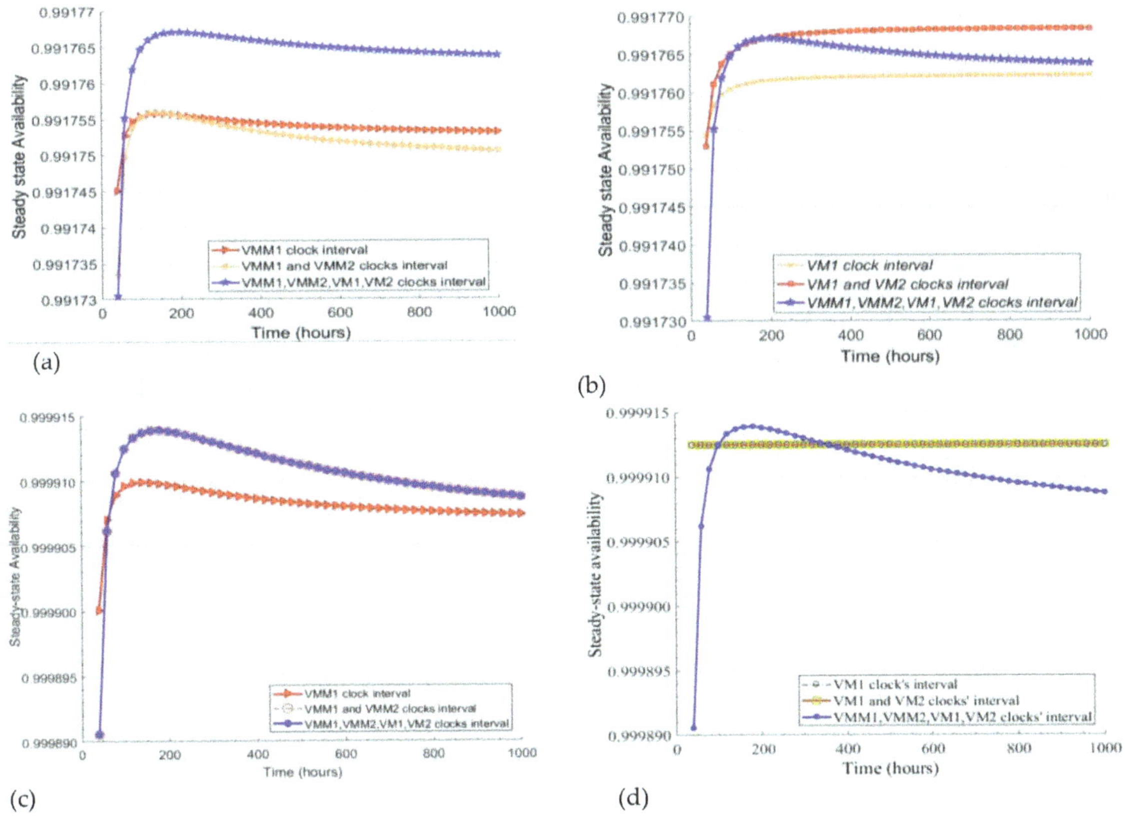

Figure 4. Sensitivity analysis of SSA of VMM and VM subsystems: (a) SSA of VM with respect to VMM clocks' intervals, (b) SSA of VM with respect to VM clocks' intervals, (c) SSA of VMM with respect to VMM clocks' intervals, and (d) SSA of VMM with respect to VM clocks' intervals.

i. **Figure 4(a)** and **(b)** shows that rejuvenation processes on VMM reduce SSA of the VM, but those on VM can improve. A proper combination of rejuvenation processes on the VMM and VM can yield an efficient impact for maintaining high values of SSA of VM.

ii. **Figure 4(c)** and **(d)** shows that there is no dependence of a VMM on its VM incorporated in the modeling of the proposed VSS yet. Also, rejuvenation implemented on both VMM subsystems of both hosts obviously gains higher SSA of VMM than it would if implemented on only one of the VMM subsystems.

4. Case study II: a DCell-based data center network

4.1. A typical DCN architecture

In this section, the DCell in consideration is expanded in size up to a network of virtualized servers complying a DCell topology. A DCell [28] is recursively constructed based on the most basic element $DCell_0$ as follows:

i. A $DCell_0$ consists of n physical servers connected to an n-port switch.

ii. A $DCell_1$ is composed of $n + 1$ $DCell_0$s. Each server of a $DCell_0$ in a $DCell_1$ has two links. One connects to its switch, the other connects to the corresponding server in another $DCell_0$, complying with a predetermined DCell routing algorithm. Consequently, every pair of $DCell_0$s in a $DCell_1$ has an exact unique link between each other.

iii. A $DCell_k$ is a level-k of $DCell_{k-1}$.

To apply the proposed modeling approach using SRN, we focus on studying a special case of DCell-based DCN at level 1 ($DCell_1$). Particularly, a cell $DCell_0$ consists of two physical servers and one shared switch. $DCell_1$ is composed of three $DCell_0$s, as shown in **Figure 5**. We assume that each server has two NICs, one for connecting to the switch in the same cell, and the other for direct connection between the server in a cell and the corresponding server in another cell, which complies with DCell network routing topology. The system architecture is detailed as follows: (i) $DCell_0[0]$ consists of switch *S0*, two hosts *H00* and *H01*, a number of VMs (n_{00} of *VM00* and n_{01} of *VM01*) on the hosts *H00* and *H01*, respectively; (ii) the description of other cells goes in the same manner.

4.2. Proposed SRN model

The SRN system model of the DCell-based DCN is presented in **Figure 6**. To simplify the modeling and to focus on sophisticated interactions between VMs and servers in a cell and in different cells of the network, we use two-state SRN models (consisting of UP and DOWN states) for physical parts of the system, including hosts and switches, as shown in **Figure 6(a)–(j)**. Initially, there is a token in the UP state for each model of a certain physical part, which is

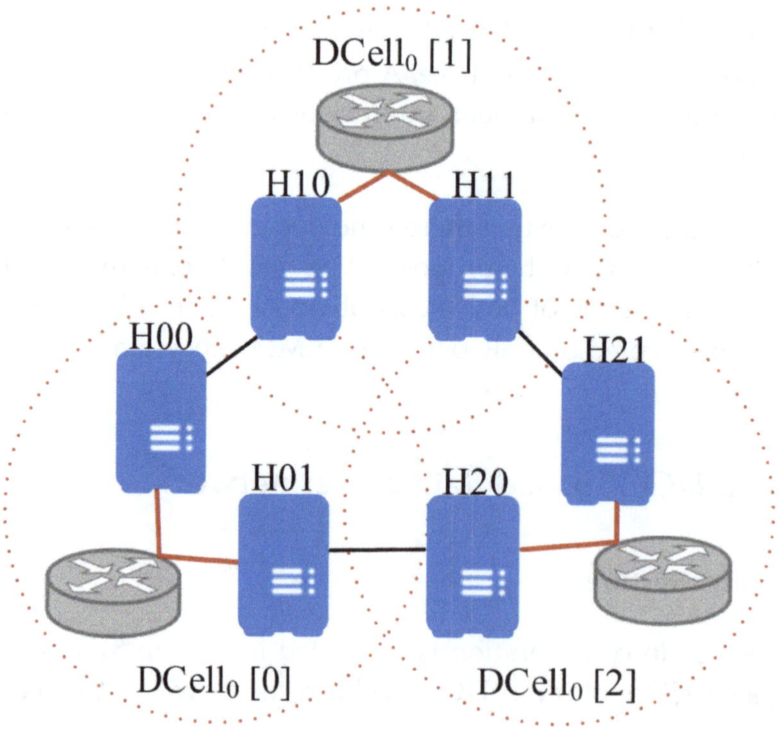

Figure 5. An architecture of a DCell-based data center network.

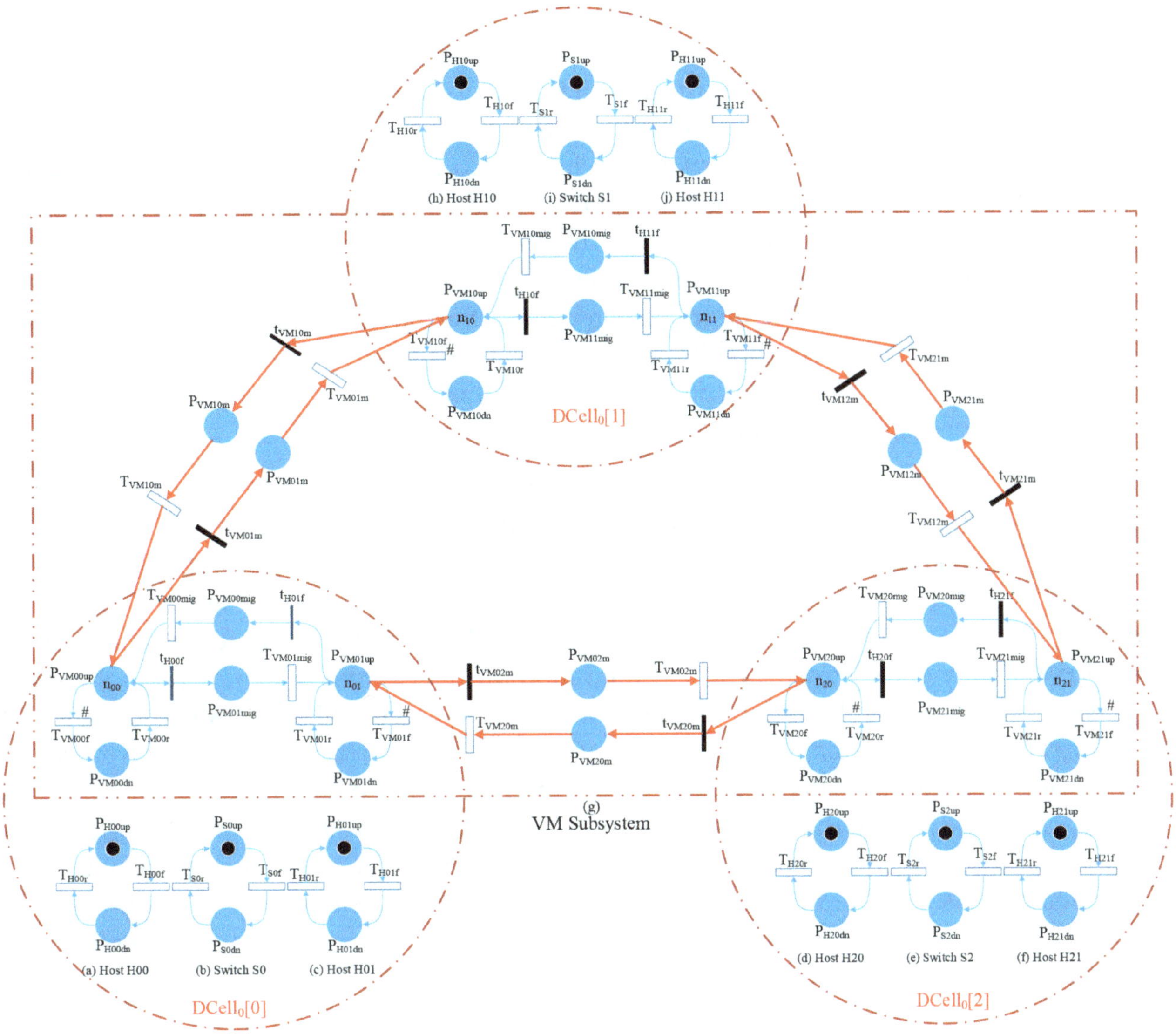

Figure 6. SRN system model of a DCell-based data center network.

depicted by a black dot which represents the initial normal working state of the physical hosts and switches. Contrary to the presented case-study of VSS in Section 3, we do not take into account the modeling of the VMM subsystem. Instead, we combine host and VMM in a unique model by considering the mean time to failure equivalent (MTTFeq) and mean time to repair equivalent (MTTReq) of the VMM subsystem as input parameters in the two-state models of hosts. Also, we simplify the modeling of the VM subsystem by using only two-state SRN models as shown in **Figure 6(g)** (VM subsystem model). There is an initial number of VMs on each host in a general case as represented by tokens in UP states. Specifically, there are n_{00} of VMs in P_{VM00up}, and n_{01} of VMs in P_{VM01up} in cell $DCell_0[0]$. In $DCell_0[1]$, the numbers of VMs initially running in a normal state on each host are n_{10} of *VM10*, and n_{11} of *VM11*, which are hosted on *H10* and *H11*, respectively. Those numbers in $DCell_0[2]$ are n_{20} of *VM20* and n_{21} of *VM21*. Unlike the SRN model of a single unit of VSS in **Figure 3**, we capture in the SRN system model the VM live migration techniques within a cell and between different cells for the sake of fault tolerance and improvement of system availability.

The VM migration is implemented between two hosts in a cell when a host in the cell experiences downtime due to a certain failure. In cell $DCell_0[0]$ for instance, the VM live migration is triggered to migrate all running VMs from the host *H00* to the host *H01* immediately when the host *H00* fails (represented by a token in P_{H00dn}). The immediate transition t_{H00f} is triggered to remove all tokens in P_{VM00up} and deposit them in $P_{VM01mig}$. As the timed transition $T_{VM01mig}$ is fired, the tokens in $P_{VM01mig}$ are removed and deposited in P_{VM01up}, representing the completion of VM live migration processes from *H00* to *H01*. If host *H01* fails (i.e., a token is placed in P_{H01dn}), the VM live migration is performed from *H01* to *H00* and is captured by the immediate transition t_{H01f} (to trigger VM live migration processes), the place $P_{VM00mig}$ (the state of a VM in migration), and the timed transition $T_{VM00mig}$ (to represent the migration processes that take time to complete). The description of VM live migration within a cell occurs in the same manner for other cells $DCell_0[1]$ and $DCell_0[2]$.

In the case of a failed switch in a cell, VM live migration is performed between two hosts in two different cells via a peer-to-peer connection. For instance, if switch *S0* fails, the connections between the two hosts *H00* and *H01* in cell $DCell_0[0]$ and the two host connections to outside users are disrupted. However, the number of *VM00* and *VM01* are still running on hosts *H00* and *H01*, respectively. It is necessary to migrate these VMs to other cells in order to enhance the overall availability of the system. The VM migration processes from cell $DCell_0[0]$ to the other two cells are triggered by the two immediate transitions t_{VM01m} (to migrate VMs from $DCell_0[0]$ to $DCell_0[1]$) and t_{VM02m} (to migrate VMs from $DCell_0[0]$ to $DCell_0[2]$). After that, the tokens in P_{VM00up} are removed and deposited in P_{VM01m} and are then deposited in P_{VM10up} in cell $DCell_0[1]$ as T_{VM01m} is fired. The transition of tokens P_{VM00up} in $DCell_0[0]$ to P_{VM10up} in cell $DCell_0[1]$ captures the migration of VM on host *H00* after a failure of switch *S0* between the two different cells. On the other side, the tokens in P_{VM01up} are removed and deposited in P_{VM02m} and are then deposited in P_{VM20up} in cell $DCell_0[2]$. This represents the migrations of VMs on host *H01* after the failure of switch *S0* from cell $DCell_0[0]$ to cell $DCell_0[2]$.

Without loss of generality, the VM live migration techniques within a cell and between two cells are described in detail as above for cell $DCell_0[0]$. These migrations apply similarly to the other cells $DCell_0[1]$ and $DCell_0[2]$.

4.3. Availability evaluation

The proposed SRN models are all implemented in SPNP. The default input parameters are listed in **Table 3**. To reduce the complexity of model analysis, we initiate only one VM on each host *H00*

Input	Description	Values	Input	Description	Values
λ_H	Host failure rate	800 hours	μ_H	Host repair rate	9.8 hours
λ_{VM}	VM failure rate	4 months	μ_{VM}	VM repair rate	30 min
λ_S	Switch failure rate	1 year	μ_S	Switch repair rate	24 hours
ω_{mig}	Network bandwidth within a $DCell_0$	1 GB/s	ω_m	Network bandwidth between two $DCell_0$s	256 Mb/s
S_{VM}	VM image size	10 GB	n_{00}, n_{01}	No. Of initial VMs in $Dcell_0[0]$	1

Table 3. Default input parameters for SRN system model of a DCN.

and *H01* in cell $DCell_0[0]$ in the default case, and there are no other VMs in the other cells. However, we also evaluate the impact of the number of VMs in the DCN on the overall system availability. In this case-study, we consider two different evaluation scenarios: (I) a standalone *DCell0* (with two hosts and one switch), and (II) the proposed three-cell DCN (as modeled above). The reward rates used to compute SSA of the two cases are defined as follows:

$$A_I = \begin{cases} 1 : \mathit{if}\ (\#P_{VM00up} + \#P_{VM01up} > 0)\&\&(\#P_{S0up} == 1) \\ 0 : \text{otherwise} \end{cases}$$

$$A_{II} = \begin{cases} 1 : \mathit{if}\{(\#P_{VM00up} + \#P_{VM01up} > 0)\&\&(\#P_{S0up} == 1)\} \\ \| \{(\#P_{VM10up} + \#P_{VM11up} > 0)\&\&(\#P_{S1up} == 1)\} \\ \| \{(\#P_{VM20up} + \#P_{VM21up} > 0)\&\&(\#P_{S2up} == 1)\} \\ 0 : \text{otherwise} \end{cases} \tag{12}$$

- Steady-state availability:
    - We first evaluate SSA and downtime of the two scenarios as shown in **Table 4**. We assume that a minute of system downtime incurs a penalty of 16,000 USD for the system owner according to the SLA *signed* with customers [29]. The results clearly show that the proposed three-cell DCN obtains much higher availability, and thus reduce downtime minutes and downtime cost penalty in a year than a standalone cell with only two physical servers.
    - We also evaluate the impact of the initial number of VMs in a DCN on the system's overall availability, as shown in **Table 5**. The results show that as we increase the initial number of VMs, the overall system availability also increases. The increased SSA in the proposed three-cell DCN is also faster than in the standalone $DCell_0$. However, if the initial number of VMs (represented by the total number of tokens in the proposed SRN system model) obtains a large value, it causes a *memory error* in computing the system availability due to the largeness problem of the SRN model.
- *Sensitivity analysis of SSA:* We observe the variation of SSA in accordance with changes in the selected input parameters, including MTTF and MTTR of hosts, VMs and switches, and VM migration rate between two hosts in a cell or in two different cells, as shown in **Figure 7**. The results show that:

Case	Description	SSA	No. of nines	Downtime (min/year)	Downtime cost (USD/year)
I	Standalone $DCell_0$	0.997240422469	2.55	1450.4	23,206,943
II	Proposed three-cell DCN	0.999950276761	4.30	26.1	418,152

Table 4. Steady-state availability and downtime cost.

n_{VM}	I		II	
	SSA	**#nines**	**SSA**	**#nines**
1	0.997064755072	2.532356	0.999773875854	3.646
2	0.997240422469	2.559157	0.999950276761	4.303
3	0.997240488479	2.559168	0.999950574780	4.306
4	0.997240519634	2.559173	0.999950839446	4.308
5	0.997240550678	2.559178	0.999951101800	4.311
6	0.997240759564	2.559210	*m.e*	*m.e*

(m.e: memory error)

Table 5. Impact of number of VMs.

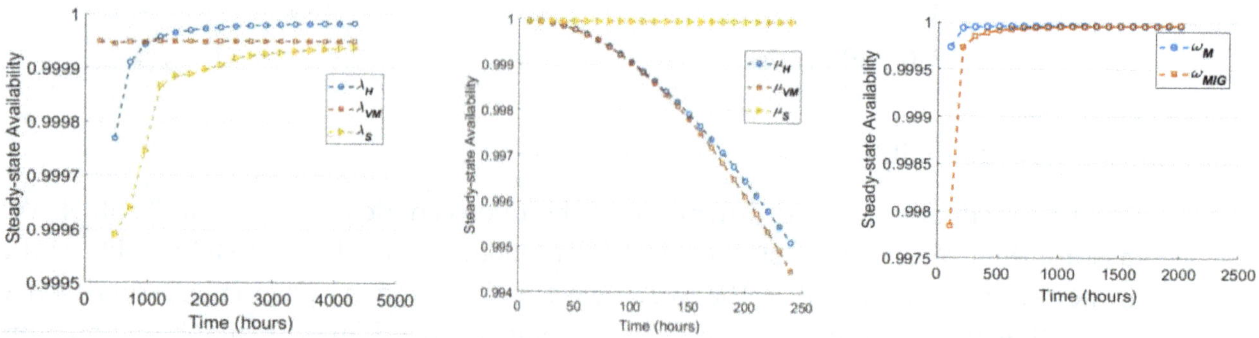

(a) (b) (c)

Figure 7. Sensitivity analysis with respect to impacting parameters. (a) MTTF, (b) MTTR, (c) VM migration rate.

- SSA is improved as we increase MTTFs and VM migration rates, and as we decrease MTTRs.
- In **Figure 7(a)**, we see that the switch is an important component of the network because its MTTF is small. Thus, the SSA clearly drops down vertically in comparison to the MTTFs of other components. Furthermore, MTTF of a host is a significant parameter in the long-run since it causes a better enhancement in the overall availability than the other MTTFs.
- In **Figure 7(b)**, we clearly find that the repair time of a switch does not affect the SSA because we perform VM migration between cells to tolerate the failures of switches. This ensures that VMs can be migrated to other cells, regardless of the failure/recovery of a certain switch. However, we can see that the recovery of a VM has a greater impact on SSA than that of a host.
- In **Figure 7(c)**, the migration rates of VMs between cells can clearly enhance SSA in comparison with those within a cell. However, the low value of the VM migration rate within a cell severely drops the system's availability.

5. Case study III: a Disaster Tolerant Data Center (DTDC)

5.1. A typical system architecture of a DTDC

This case-study considers disaster tolerance of cloud computing in a DCS. The system is composed of two different DCs (DC1 and DC2), which are geographically located in two distant regions, as shown in **Figure 8**. In each DC, we place a VSS of two physical servers (H1 and H2 in DC1, and H3 and H4 in DC2). All physical machines are assumed to be identical. Each server is initially capable of running a VM (VM1~VM4 runs on H1~H4, respectively). Shared network attached storage (NAS) is equipped in each DC to provide distributed storage and a VM migration mechanism between two hosts in the same DC. To implement disaster tolerance and recovery strategies between DCs, a back-up server is incorporated to provide VM data backup. The back-up server allows periodic synchronization of VM data between DCs. This allows the most-updated VM data to be recovered onto an operational DC after a disaster strikes on another DC.

Furthermore, to enhance the system's overall availability, we use the (active-standby) fail-over technique and VM switching mechanism. Specifically, when a VM on a certain host fails, a standby VM on the same host wakes up and takes over the operations of the failed VM. If there is no standby VM on the same host, the standby VM on the remaining host goes up and takes place on the failed host.

If a host in a DC fails, its VMs in the standby state are switched on in order to load onto the remaining host. Various VM migration mechanisms are also taken into account in this system. VM live-migration is performed between two hosts in a DC when one of the hosts fails. VM migration between two DCs is triggered when a DC undergoes a system failure when two hosts enter a downtime period simultaneously. When a disaster devastates a DC, VM migration between the back-up server (in a safe zone) and the remaining operational DC is implemented as a means of disaster recovery.

5.2. Availability modeling of a DTDC

The SRN system model for availability quantification of the studied DTDC is shown in **Figure 9**. We use simplified two-state SRN models (UP and DOWN) to capture general failure and recovery behaviors of physical parts in the system, including the physical hosts H1–H4 (**Figure 9(a), (b), (j),**

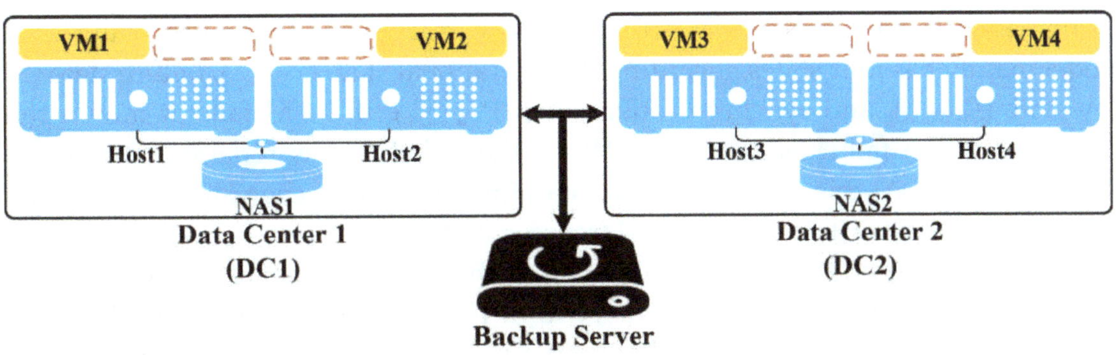

Figure 8. A conceptual architecture of a disaster tolerant data center system.

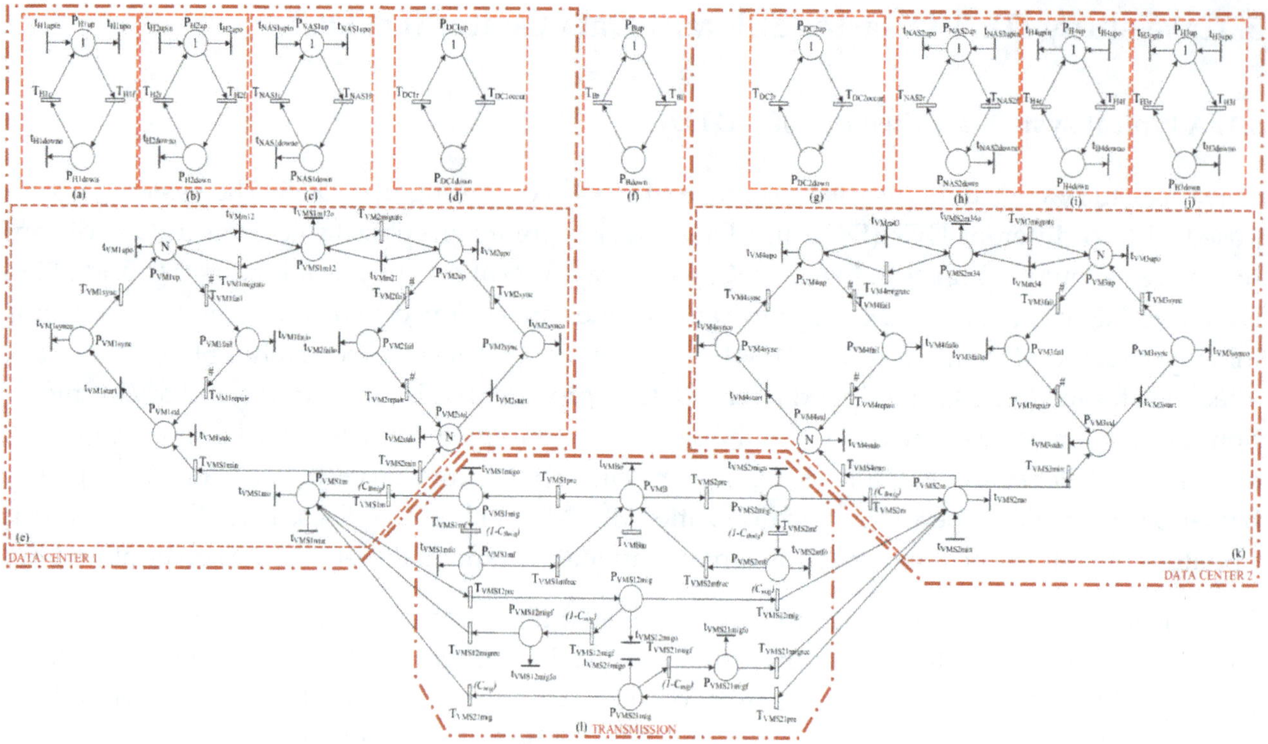

Figure 9. SRN system model of a disaster tolerant data center.

and **(i)**, respectively), NAS1 in DC1, and NAS2 in DC2 (**Figure 9(c)** and **(h)**, respectively). We use immediate transitions t_{Hupo}, t_{Hdowno}, t_{NASupo}, and $t_{NASdowno}$ to remove tokens in the up and down places of the host and NAS models in order to represent the entire operational termination of a DC when a disaster strikes. When the disaster passes and the reconstructed DC starts a new operational cycle, the immediate transitions t_{Hupin} and $t_{NASupin}$ are used to deposit new tokens in the up states of the host and NAS models. The occurrence of a disaster at a site is also represented by using a two-state model as shown in **Figure 9(d)** and **(g)** for the occurrence of a disaster at DC1 and DC2, respectively. The two-state SRN model in **Figure 9(f)** captures the operational and failure states of the back-up server.

The modeling of VM subsystems in DC1 and DC2 are shown in **Figure 9(e)** and **(k)**, respectively. Since we initially assume that all hosts and VMs are identical, the modeling of the two DCs is also identical. The model initializes N tokens in P_{VM1up}, and the other N tokens in P_{VM2std} represent N operational VMs with their N standby VMs at the beginning. Each VM sub-model mainly has four states, including the operational state (P_{VMup}), failure state (P_{VMfail}), standby state (P_{VMstd}), and synchronization state (P_{VMsync}). If a VM fails, it moves from the upstate P_{VMup} to the failure state P_{VMfail}. When the failed VM is repaired, it moves to the standby state P_{VMstd}. At this point, the active-standby fail-over mechanism of VMs is captured as follows. When a VM fails, a standby VM (represented by a token in P_{VMstd}) on the same host (before the disaster) or on the remaining host (after the disaster) transits to P_{VMsync} in order to synchronize the most-updated data on the NAS of that DC corresponding to the previously failed VM. It then goes up to P_{VMup} and takes the place of the failed VM. Dependence marks are placed near timed transitions T_{VMfail} and $T_{VMrepair}$ to represent the competition between failure and repair of VMs on the same host. The VM live-migration technique is triggered as a host fails, which is captured by an immediate transition t_{VMm}, a place P_{VMSm}, and a timed transition $T_{VMmigrate}$. For instance, when host H1 fails, the VM live-migration is triggered to migrate running VMs from the failed H1 to the running H2. Thus, t_{VMm12} is triggered to fire. A number of tokens in P_{VM1up} are removed and deposited at $P_{VMS1m12}$ as it waits for migration. The timed transition $T_{VM2migrate}$ is then fired to depict the migration process of VMs onto host H2. The tokens in $P_{VMS1m12}$ are removed and deposited in P_{VM2up}. The reversed migration from host H2 to H1 is captured by t_{VMm21}, $P_{VMS1m12}$, and $T_{VM1migrate}$ in the same manner. The places P_{VMS1m} and P_{VMS2m} represent the storage of VMs on NAS1 and NAS2. When the two hosts in a DC enter downtime, all tokens in the VM sub-models of VM1 and VM2 are removed by immediate transitions t_{VMupo}, $t_{VMfailo}$, t_{VMstdo}, and $t_{VMsynco}$ (attached to four main states of VM sub-models) and deposited in P_{VMS1m} via $t_{VMS1min}$. However, if a disaster strikes, the all tokens are removed from the places in the VM sub-models via the out-going immediate transitions t_{VMupo}, $t_{VMfailo}$, t_{VMstdo}, $t_{VMsynco}$, t_{VMSmo}, and t_{VMSmo}. As the failed data center is reconstructed, a pre-defined number of VMs are created on the NAS, which is captured by depositing tokens in P_{VMSm} via t_{VMSmin}. The VMs are then assigned to hosts via the time transition T_{VMSmin}.

The VM migration techniques between the two DCs, and between the backup server and the two DCs, are modeled in **Figure 9(l)**. The place P_{VMB} represents the storage of VMs in the back-up server. When a DC is destroyed due to a disaster, its VMs are stored in the back-up server and represented by creating new tokens in P_{VMB} via the timed transition T_{VMBin}. When there is a remaining DC in its operational state, the tokens in P_{VMB} are transmitted to the corresponding P_{VMSmig} via the timed transition T_{VMSpre}. The tokens are then deposited in P_{VMSm} via the timed transition T_{VMSm} of the respective DC model with an imperfect coverage factor C_{Bmig}. If this process fails with coverage factor (1-C_{Bmig}), the tokens are moved to P_{VMS2mf} via T_{VMSmf} and returned to P_{VMB} via $T_{VMSmfrec}$. This transition of tokens captures the VM migration from the back-up server to the operational DC. In the case when the back-up server fails, the immediate transitions t_{VMBo}, $t_{VMSmigo}$, and t_{VMSmfo} remove all tokens in P_{VMB}, P_{VMSmig}, and P_{VMSmf} to represent the loss of VM image files on the back-up server. The VMs will be created on the back-up server as soon as it is recovered. The VM migration between two DCs is triggered when two hosts in a DC enter downtime simultaneously. In this case, we propose the two hosts H1 and H2 in DC1 also stay in a downtime period simultaneously. A number of VMs on DC1 are still stored in NAS1, represented by tokens in P_{VMS1m}. Thus, it is necessary to migrate these VMs onto the running DC2. The tokens are then transmitted to $P_{VMS12mig}$ after a pre-migration process ($T_{VMS12pre}$). The VM migration process is finalized with an imperfect coverage factor C_{mig} as the transition $T_{VMS12mig}$ is fired. If this migration process fails with coverage factor (1-C_{mig}), the tokens are moved to $P_{VMS12migfo}$ and returned to NAS1 in the original DC1 via $T_{VMS12migrec}$. The VM migration from DC2 to DC1 is performed similarly and captured by the places $P_{VMS21mig}$, $P_{VMS21migf}$, the timed transition $T_{VMS21pre}$, $T_{VMS21mig}$ (with imperfect coverage factor C_{mig}), $T_{VMS21migf}$ (with coverage factor 1-C_{mig}), and $T_{VMS21migrec}$.

5.3. Availability evaluation

The SRN system model is implemented in SPNP. Default input parameter values are shown in **Table 6**. We assume that the number of VMs on a host is only one in order to reduce complexity in model computation and analysis.

Input	Description	Assigned transitions	Values
λ_{Hf}	Host failure rate	T_{H1f}, T_{H2f}, T_{H3f}, T_{H4f}	800 hours
μ_{Hr}	Host recovery rate	T_{H1r}, T_{H2r}, T_{H3r}, T_{H4r}	9.8 hours
λ_{NASf}	NAS failure rate	T_{NAS1f}, T_{NAS2f}	45 years
μ_{NASr}	NAS recovery rate	T_{NAS1r}, T_{NAS2r}	4 hours
$\lambda_{DCoccur}$	Time to disaster occurrence at a DC	$T_{DC1occur}$, $T_{DC2occur}$	100 years
μ_{DCr}	DC recovery rate after a disaster	T_{DC1r}, T_{DC2r}	1 year
λ_{Bf}	Backup DC failure rate	T_{Bf}	50,000 hours
μ_{Br}	Backup DC recovery rate	T_{Br}	30 min
λ_{VMfail}	VM failure rate	$T_{VM1fail}$, $T_{VM2fail}$, $T_{VM3fail}$, $T_{VM4fail}$	4 months
$\mu_{VMrepair}$	VM repair rate	$T_{VM1repair}$, $T_{VM2repair}$, $T_{VM3repair}$, $T_{VM4repair}$	30 min
δ_{VMsync}	VM synchronization rate	$T_{VM1sync}$, $T_{VM2sync}$, $T_{VM3sync}$, $T_{VM4sync}$	5 min
$\omega_{VMmigrate}$	VM migration rate between hosts	$T_{VM1migrate}$, $T_{VM2migrate}$, $T_{VM3migrate}$, $T_{VM4migrate}$	5s
γ_{VMSmin}	VM loading rate into a host	$T_{VMS1min1}$, $T_{VMS1min2}$, $T_{VMS2min3}$, $T_{VMS2min4}$	1 s
η_{VMSpre}	VM pre-migration rate between DCs and backup server	$T_{VMS12pre}$, $T_{VMS21pre}$, $T_{VMS1pre}$, $T_{VMS2pre}$	5 min
$\theta_{VMSmigrec}$	VM return rate to NAS after a migration failure	$T_{VMS12migrec}$, $T_{VMS21migrec}$	1 min
$\theta_{VMSmfsync}$	VM synchronization rate with backup DC after a migration failure	$T_{VMS1mfsync}$, $T_{VMS2mfsync}$	1 min
C_{Bmig}	Imperfect factor of VM migration from backup DC		0.95
C_{mig}	Imperfect factor of VM migration between DCs		0.85
N	Number of VMs in a host		1
S_{VM}	Size of VM image and related data		4GB
ω_{NET}	Network speed		20 MB/s

Table 6. Default input parameters.

- *Steady state availability*: We evaluate the availability of the DTDC in seven operational scenarios by varying imperfect VM migration coverage factors between the backup server and the DCs and disaster occurrence frequency as follows: (I) The system of two standalone DCs without DT confronts disasters at the mean time to occurrence of 100 years (default value); (II) The system with default parameters; (III-V) The network connection has a high probability of failure (i.e., low probability of success in VM migration processes) and the system is planted in an area with mean disaster time set alternatively to 100, 200, and 300 years; (VI-VIII) In contrast to cases (III)-(V), the migration between distant parts may succeed with high probability and the DCs location experiences disasters with mean time to occurrence also set to 100, 200, and 300 years. The results of SSA and downtime evaluation are shown in **Table 7** such that following criteria are satisfied:

Case	C_{Bmig}	C_{mig}	$\lambda_{DCoccur}$	SSA	No. of nines	Downtime (min/year)	Downtime cost (USD/year)
I	x	X	100 years	0.989455392105	1.98	5542.2	8,675,934.6
II	0.95	0.85	100 years	0.999843164703	3.80	82.4	1,318,922.1
III	0.1	0.1	100 years	0.998942162067	2.98	556.0	8,895,993.9
IV	0.1	0.1	200 years	0.999635096345	3.44	191.8	3,068,693.8
V	0.1	0.1	300 years	0.999795681447	3.69	107.4	1,718,237.3
VI	0.9	0.9	100 years	0.999841085616	3.80	83.5	1,336,406.4
VII	0.9	0.9	200 years	0.999946639371	4.27	28.0	448,741.5
VIII	0.9	0.9	300 years	0.999968676113	4.50	16.5	263,421.4

Table 7. SSA and downtime analyses.

- The safer DCs locations (longer frequency of disaster occurrence) results in a higher system SSA.
- DCs should be placed in isolated areas to avoid any severe damage from disastrous events, even though the network connection between distant parts of the system might deal with more failure during VM migration processes.
- Higher SSA values are obtained with more reliable network connections, i.e. for network connections that can guarantee a higher success rate for transmission between distant parts of the system.

- *Sensitivity analysis:* As shown in **Figure 10**, we analyzed the sensitivity of the system's SSA with respect to different parameters, including imperfect coverage factors of VM migration (C_{Bmig} and C_{mig}), time to disaster occurrences ($\lambda_{DCoccur}$), VM image size (S_{VM}), and network bandwidth (ω_{NET}). The impact of S_{VM} and ω_{NET} is shown in **Figure 10(f)**. The

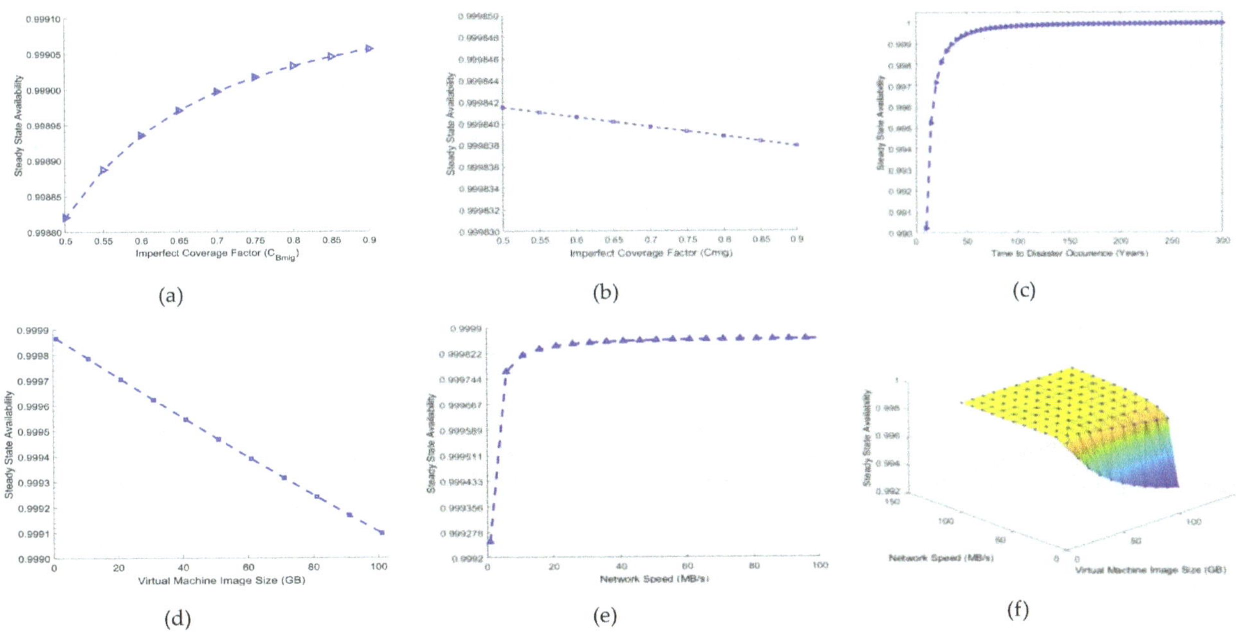

Figure 10. Sensitivity analysis of a DTDC steady state availability: (a) C_{Bmig}, (b) C_{mig}, (c) $\lambda_{DCoccur}$, (d) S_{VM}, (e) ω_{NET}, (f) ω_{NET}, S_{VM}.

results show that: (i) the disaster tolerance solution with a back-up center would improve SSA, even when connections between the back-up center with DCs incur imperfections in VM migration processes; (ii) imperfections in the VM migration processes between DCs slightly impact SSA when it increases; (iii) the system's SSA is improved vastly if DCs are located in safe areas with lower disaster occurrence frequency; (iv) larger VMs can reduce the overall availability of the system; (v) a faster network connection between distant locations can actually boost the system's availability, especially for network speeds ranging in 0-20 Mb/s, if the speed increases much higher, the effect is not much different from the default parameters; (vi) the variation of both ($\omega_{NET,}$ S_{VM}) confirms the fact that higher network speed and smaller VM sizes result in apparently higher SSA, whereas slower network and larger VMs severely reduce the system's availability.

6. Conclusion(s)

This chapter presented a set of availability models based on stochastic reward net for comprehensive system availability evaluation in data center systems. The data center systems scale during evaluation was increased from a system of two virtualized servers (considered as a unit block in data centers) in Section 3, to a typical network of virtualized servers complying with a DCell topology in Section 4. Finally, the evaluated data centers are scaled up to a two-site data center for disaster tolerance with a back-up center. A variety of fault and disaster tolerant techniques were incorporated in the systems in order to achieve high availability. The systems were evaluated under various case studies with regards to different metrics of interest, including steady state availability and its sensitivity with respect to a number of impac factors. The analysis results show comprehensive system behaviors and improved availability in accordance with incorporated techniques in the data center systems.

Acknowledgements

This research was supported by the Ministry of Science, ICT (MSIT), Korea, under the Information Technology Research Center (ITRC) support program (IITP-2018-2016-0-00465) supervised by the Institute for Information & communications Technology Promotion (IITP).

Author details

Tuan Anh Nguyen[1,2]*, Dugki Min[1] and Eunmi Choi[3]

*Address all correspondence to: anhnt2407@gmail.com

1 Office of Research, University-Industry Cooperation Foundation, Konkuk University, Seoul, South Korea

2 Department of Computer Engineering, Konkuk University, Seoul, South Korea

3 School of Management Information Systems, Kookmin University, Seoul, South Korea

References

[1] P. Insitute. 2016 Cost of Data Center Outages. Ponemon Inst; 2016

[2] Sony M, Mariappan V, Kamat V. Stochastic modelling of failure interaction: Markov model versus discrete event simulation. International Journal of Advanced Operations Management. 2011;**3**(1):1

[3] Szczerbicka H, Trivedi KS, Choudhary PK. Discrete event simulation with application to computer communication systems performance. In: Reis R, editor. Information Technology. Boston: Kluwer Academic Publishers; 2004. pp. 271-304

[4] Trivedi KS, Kim DS, Roy A, Medhi D. Dependability and security models. In: Proc. 2009 7th Int. Work. Des. Reliab. Commun. Networks, DRCN 2009; Oct. 2009. pp. 11-20

[5] Han S, Nashville T. Multidisciplinary System Reliability Analysis. NASA Contract Report. NASA CR-210969. Jun 2001. Available from: http://gltrs.grc.nasa.gov/reports/2001/CR-2001-210969.pdf

[6] Cao Y, Sun H, Trivedi KS, Han JJ. System availability with non-exponentially distributed outages. IEEE Transactions on Reliability. Jun 2002;**51**(2):193-198

[7] Trivedi KS, Kim DS, Ghosh R. System availability assessment using stochastic models. Applied Stochastic Models in Business and Industry. Mar 2013;**29**(2):94-109

[8] Nguyen TA, Min D, Choi E. A comprehensive evaluation of availability and operational cost for a virtualized server system using stochastic reward nets. The Journal of Supercomputing. Aug 2017:1-55

[9] Han K, Nguyen TA, Min D, Choi EM. An evaluation of availability, reliability and power consumption for a SDN infrastructure using stochastic reward net. In: Park JH, Pan Y, Yi G, Loia V, editors. Advances in Computer Science and Ubiquitous Computing: CSA-CUTE 2016; Singapore: Springer Singapore. 2017. pp. 637-648

[10] Raei H, Yazdani N. Performability analysis of cloudlet in mobile cloud computing. Inf. Sci. (Ny). 2017

[11] Nguyen TA, Eom T, An S, Park JS, Hong JB, Kim DS. Availability modeling and analysis for software defined networks. In: 2015 IEEE 21st Pacific Rim International Symposium on Dependable Computing (PRDC); 2015 April. pp. 159-168

[12] Dantas J, Matos R, Araujo J, Maciel P. Eucalyptus-based private clouds: Availability modeling and comparison to the cost of a public cloud. Computing. Nov 2015;**97**(11): 1121-1140

[13] Andrade E, Nogueira B, Matos R, Callou G, Maciel P. Availability modeling and analysis of a disaster-recovery-as-a-service solution. Computing. Feb 2017:1-26

[14] Raei H, Yazdani N, Shojaee R. Modeling and performance analysis of cloudlet in mobile cloud computing. Performance Evaluation. 2017;**107**:34-53

[15] Patel P, Ranabahu A, Sheth A. Service Level Agreement in Cloud Computing. Kno.e.sis Publications. The Ohio Center of Excellence in Knowledge Enabled Computing (Kno.e.sis). 2009. Available from: http://corescholar.libraries.wright.edu/knoesis/78

[16] Garg SK, Toosi AN, Gopalaiyengar SK, Buyya R. SLA-based virtual machine management for heterogeneous workloads in a cloud datacenter. Journal of Network and Computer Applications. Aug 2014;**45**:108-120

[17] Nanda S, Chiueh T. A Survey of Virtualization Technologies. SUNY; 2005

[18] Daniels J. Server virtualization architecture and implementation. Crossroads. Sep. 2009; **16**(1):8-12

[19] Ameen RY, Hamo AY. Survey of server virtualization. International Journal of Computer Science and Information Security. Apr 2013;**11**(3):65-74

[20] Kim DS, Machida F, Trivedi KS. Availability modeling and analysis of a virtualized system. In: 2009 15th IEEE Pacific Rim International Symposium on Dependable Computing, PRDC 2009; **2009**

[21] Smith WE, Trivedi KS, Tomek LA, Ackaret J. Availability analysis of blade server systems. IBM Systems Journal. 2008;**47**(4):621-640

[22] Grottke M, Nikora AP, Trivedi KS. An empirical investigation of fault types in space mission system software. In: Proceedings of 2010 IEEE/IFIP International Conference on Dependable Systems & Networks (DSN); 2010. pp. 447-456

[23] Machida F, Xiang J, Tadano K, Maeno Y. Combined server rejuvenation in a virtualized data center. In: 2012 9th International Conference on Ubiquitous Intelligence and Computing and 9th International Conference on Autonomic and Trusted Computing; 2012. pp. 486-493

[24] Cui L, Li B, Li J, Hardy J, Liu L. Software aging in virtualized environments: Detection and prediction. In: Proceedings of 2012 IEEE 18th International Conference on Parallel and Distributed Systems; 2012. pp. 718-719

[25] Longo F, Ghosh R, Naik VK, Trivedi KS. A scalable availability model for Infrastructure-as-a-Service cloud. In: Proceedings of 2011 IEEE/IFIP 41st International Conference on Dependable Systems & Networks (DSN); 2011. pp. 335-346

[26] Ciardo G, Muppala J, Trivedi KS. SPNP: Stochastic petri net package. In: Proc. Third Int. Work. Petri Nets Perform. Model. PNPM89; 1989. pp. 142-151

[27] Machida F, Kim DS, Trivedi KS. Modeling and analysis of software rejuvenation in a server virtualized system with live VM migration. Performance Evaluation. 2013;**70**(3):212-230

[28] Guo C, Wu H, Tan K, Shi L, Zhang Y, Lu S. Dcell: A scalable and fault-tolerant network structure for data centers. In: Proceedings of the ACM SIGCOMM 2008 Conference on Data Communication—SIGCOMM '08. Vol. 38(4). 2008. p. 75

[29] Stansberry M. 2013 Data Center Industry Survey. Uptime Institute, LLC; 2013

Permissions

All chapters in this book were first published in TSAA, by InTech Open; hereby published with permission under the Creative Commons Attribution License or equivalent. Every chapter published in this book has been scrutinized by our experts. Their significance has been extensively debated. The topics covered herein carry significant findings which will fuel the growth of the discipline. They may even be implemented as practical applications or may be referred to as a beginning point for another development.

The contributors of this book come from diverse backgrounds, making this book a truly international effort. This book will bring forth new frontiers with its revolutionizing research information and detailed analysis of the nascent developments around the world.

We would like to thank all the contributing authors for lending their expertise to make the book truly unique. They have played a crucial role in the development of this book. Without their invaluable contributions this book wouldn't have been possible. They have made vital efforts to compile up to date information on the varied aspects of this subject to make this book a valuable addition to the collection of many professionals and students.

This book was conceptualized with the vision of imparting up-to-date information and advanced data in this field. To ensure the same, a matchless editorial board was set up. Every individual on the board went through rigorous rounds of assessment to prove their worth. After which they invested a large part of their time researching and compiling the most relevant data for our readers.

The editorial board has been involved in producing this book since its inception. They have spent rigorous hours researching and exploring the diverse topics which have resulted in the successful publishing of this book. They have passed on their knowledge of decades through this book. To expedite this challenging task, the publisher supported the team at every step. A small team of assistant editors was also appointed to further simplify the editing procedure and attain best results for the readers.

Apart from the editorial board, the designing team has also invested a significant amount of their time in understanding the subject and creating the most relevant covers. They scrutinized every image to scout for the most suitable representation of the subject and create an appropriate cover for the book.

The publishing team has been an ardent support to the editorial, designing and production team. Their endless efforts to recruit the best for this project, has resulted in the accomplishment of this book. They are a veteran in the field of academics and their pool of knowledge is as vast as their experience in printing. Their expertise and guidance has proved useful at every step. Their uncompromising quality standards have made this book an exceptional effort. Their encouragement from time to time has been an inspiration for everyone.

The publisher and the editorial board hope that this book will prove to be a valuable piece of knowledge for researchers, students, practitioners and scholars across the globe.

List of Contributors

Wlodek M. Zuberek
Department of Computer Science, Memorial University, St. John's, NL, Canada

Dariusz Strzeciwilk
Department of Applied Informatics, University of Life Sciences, Warszawa, Poland

Mahmoud Ghofrani and Anthony Suherli
Electrical Engineering, Engineering and Mathematics Division, School of STEM, University of Washington Bothell, USA

Ireneusz Czmoch
Department of Structural Mechanics and Computer Aided Engineering, Faculty of Civil Engineering, Warsaw University of Technology, Warsaw, Poland

Erica Teixeira Gomes de Sousa and Fernando Antonio Aires Lins
Department of Statistics and Informatics, Federal Rural University of Pernambuco, Brazil

Lena Feinbube, Lukas Pirl and Andreas Polze
Hasso-Plattner-Institute, University Potsdam, Germany

Ali Jannesari
Department of Computer Science, Iowa State University, USA

Panagiotis Sismanis
SIDENOR SA, Athens, Greece

Simon Volpert, Frank Griesinger and Jörg Domaschka
Institute of Information Resource Management, Ulm University, Ulm, Germany

Urbanus F Melkior, Josef Tlustý and Zdeněk Müller
Department of Electrical Power Engineering, Czech Technical University, Prague, Czech Republic

Taizhi Liu, Chang-Chih Chen and Linda Milor
Georgia Institute of Technology, Georgia

Tuan Anh Nguyen
Office of Research, University-Industry Cooperation Foundation, Konkuk University, Seoul, South Korea
Department of Computer Engineering, Konkuk University, Seoul, South Korea

Dugki Min
Office of Research, University-Industry Cooperation Foundation, Konkuk University, Seoul, South Korea

Eunmi Choi
School of Management Information Systems, Kookmin University, Seoul, South Korea

Index

S

T

V

W

www.ingramcontent.com/pod-product-compliance
Lightning Source LLC
LaVergne TN
LVHW060604210926
848485LV00005B/1366